*Growth, maturation, and body composition* documents one of the most remarkable and significant long-term studies in the field of human biology. The Fels Longitudinal Study is the longest, largest and most productive serial study of human growth, maturation and body composition. This book shows how data collected from more than 1000 participants during the past 60 years have been analysed to test a wide range of hypotheses, and describes how the findings have led to the development of improved research methods. With more than 1000 specialized publications of Fels data to date, the present book provides a unique overview of this fascinating research program, which will be of interest to a wide range of researchers, including those in the fields of physical anthropology, nutrition science, pediatrics, gerontology, epidemiology, endocrinology, human genetics, as well as statistics.

*Cambridge Studies in Biological Anthropology 9*

Growth, maturation, and body composition
*the Fels Longitudinal Study 1929–1991*

*Cambridge Studies in Biological Anthropology*

*Series Editors*

G.W. Lasker
Department of Anatomy, Wayne State University,
Detroit, Michigan, USA

C.G.N. Mascie-Taylor
Department of Biological Anthropology,
University of Cambridge

D.F. Roberts
Department of Human Genetics,
University of Newcastle-upon-Tyne

R.A. Foley
Department of Biological Anthropology,
University of Cambridge

*Also in the series*

G.W. Lasker *Surnames and Genetic Structure*
C.G.N. Mascie-Taylor and G.W. Lasker (editors) *Biological Aspects of Human Migration*
Barry Bogin *Patterns of Human Growth*
Julius A. Kieser *Human Adult Odontometrics – The study of variation in adult tooth size*
J.E. Lindsay Carter and Barbara Honeyman Heath *Somatotyping – Development and applications*
Roy J. Shephard *Body Composition in Biological Anthropology*
Ashley H. Robins *Biological Perspectives on Human Pigmentation*
C.G.N. Mascie-Taylor and G.W. Lasker (editors) *Applications of Biological Anthropology to Human Affairs*

# Growth, maturation, and body composition

## the Fels Longitudinal Study
## 1929–1991

ALEX F. ROCHE

*Professor of Community Health and Pediatrics, Wright State University,*
*Dayton, Ohio, USA*

CAMBRIDGE
UNIVERSITY PRESS

CAMBRIDGE UNIVERSITY PRESS
Cambridge, New York, Melbourne, Madrid, Cape Town, Singapore, São Paulo

Cambridge University Press
The Edinburgh Building, Cambridge CB2 8RU, UK

Published in the United States of America by Cambridge University Press, New York

www.cambridge.org
Information on this title: www.cambridge.org/9780521374491

First published 1992
This digitally printed version 2008

A catalogue record for this publication is available from the British Library

ISBN 978-0-521-37449-1 hardback
ISBN 978-0-521-05512-3 paperback

*To*
*Stanley Marion Garn*

*with respect and admiration for his unlimited energy and vision*

# Contents

ix

# *Preface*

I am fortunate to have been able to write this book. This good fortune began with my appointment to the Fels Staff 23 years ago. It has taken me all that time to understand the complexities of the Fels Longitudinal Study. Writing this book about the study has been a pleasure. I trust the reader will find it pleasant also. Quoting Daniel Defoe: 'If this work is not both pleasant and profitable to the reader ... the fault ... cannot be any deficiency in the subject.'

Many helped. Some added to the quality of the study, particularly the generous participants and their relatives, outstanding collaborators and consultants, efficient secretaries and dedicated research assistants. Particular thanks are due to Ruth Bean and Lois Croutwater who, during a joint span of more than 56 years of extraordinary effort, have organized the examinations and maintained contact with the participants. Doctors Lester Sontag and Frank Falkner, the past Directors, provided effective leadership from 1929 to 1979. The Fels Longitudinal Study, supported in its early years by the Samuel S. Fels Fund, continues as part of the Division of Human Biology of the Department of Community Health at Wright State University. This division receives enthusiastic support from the Departmental Chairman (Dr Robert Reece), the School of Medicine and the central administration of Wright State University. Continuation of this support is important because the story is not complete. The significance of the Fels Longitudinal Study increases as the data base enlarges, the serial records become longer, and new techniques are introduced. Additionally, the focus of the Study is shifting rapidly to more applied areas.

Many helped write this book, but they are not to blame for its defects. Jean Bolin prepared the illustrations, and Joan Hunter typed all the early drafts. The later drafts, typed by the helpful staff of the Word Processing Center in the School of Medicine, were read by Rick Baumgartner, Cameron Chumlea, Shumei Guo, Roger Siervogel, and my wife, Eileen. Nancy Kern of the Health Sciences Library and Cheryl Caddell were indefatigable in their search for sources and Jane Smith copy-edited the manuscript thoroughly and pleasantly. The Samuel S. Fels Fund provided encouragement and generous financial assistance. Last in this sequence, but not least in importance, are Dr Alan Crowden and Dr Sara Trevitt of the Cambridge University Press who waited patiently for the delayed manuscript and guided it through all stages of the publication process in a charming and effective manner. To all these friends, I am most grateful.

Alex F. Roche

# *Abbreviations*

BD = body density (g/cc)
DNA = deoxyribonucleic acid
LPCA = longitudinal principal component analysis
P = probability
PHV = age at peak height velocity (years)
PTC = phenylthiocarbamide
R = resistance (ohms)
SD = standard deviation
S = stature (cm)
W = weight (kg)
$X_c$ = reactance (ohms)
Z = impedance (ohms)
$\Phi$ = phase angle (degrees)

# 1    *Introduction*

> What has been accomplished is only an earnest of what shall be done in
> the future. Upon our heels a fresh perfection must tread, born of us, fated
> to excel us. We have but served and have but seen a beginning. Personally,
> I feel deeply grateful to have been permitted to join in this noble work and
> to have been united in it with men (and women) of such high and human
> ideas.
>
> *William Osler* (1849–1919)

The above quotation from the writings of Sir William Osler is fully
appropriate. It has been an honor for me to work in the Fels Longitudinal
Study. Indeed, all the members of the Fels staff are fully conscious of their
debt to those who laid the foundations for the studies that are summarized
in this volume. Although much has been achieved, all realize that those
who follow will greatly extend our present limited horizons.

This book describes the progress that has been made during the first 60
years of the Fels Longitudinal Study of Growth, Maturation and Body
Composition. The remarkable nature of the study justifies this volume.
Very few, if any, investigations of human beings are so longlived. Despite
its longevity, the Fels Longitudinal Study continually becomes more
vigorous and active. Few studies have been responsible for equally
important research related to serial changes in physical growth and
maturation and body composition in normal individuals.

The Fels Research Institute was founded in 1929 with a single complex
research project that came to be called the Fels Longitudinal Study. As the
name implies, this was a serial study and it was multidisciplinary. Between
1927 and 1932 several such studies were founded in the US, which in
addition to the Fels Longitudinal Study, included the Bolton–Brush Study
at Western Reserve University (Cleveland, Ohio), the Berkeley Growth
Study, Guidance Study, and Oakland Growth Study at the Institute of
Human Development of the University of California (Berkeley), the Child

1

Research Council Study at the University of Colorado (Denver), and the Harvard School of Public Health Growth Study (Boston, Massachusetts).

This sudden rush to begin serial multidisciplinary studies of child growth and development between 1927 and 1932 sprang, in part, from desires to shield children from the worst effects of the Great Depression. These same concerns led to the 1933 White House Conference on Child Health and Protection which recommended that such studies be undertaken. It was recognized that further knowledge was required to determine both the effects of the Great Depression and the possible influence of programs intended to mitigate these effects.

Serial data for individuals were recorded in each of these studies but, generally, the data were analyzed cross-sectionally. Each study had its area of concentration and outstanding expertise. The Bolton–Brush Study was responsible for major advances in methods for the study of craniofacial growth (Broadbent, 1931; Broadbent, Broadbent & Golden, 1975), and of maturation (Greulich & Pyle, 1959), and useful information was provided about growth changes in various anthropometric dimensions (Simmons, 1944). This study is scarcely active, although occasional use is made of the records which are intact but not in an electronic format.

The three studies at Berkeley (Berkeley Growth Study, Guidance Study, Oakland Growth Study) were responsible for important research contributions in child growth, psychology and sociology, including advances in methods for the analysis of serial data and the measurement of psychological development. The results of these studies have been summarized by Jones *et al.* (1971). The Berkeley group of studies is now inactive except for occasional sociological studies of participants in adulthood.

The Child Research Council Study contributed valuable information about diet, tissue growth, electrocardiography and basal metabolic rate, although few of the analyses were serial. Data collection in this study ceased in 1968. Many of the recorded data are stored electronically and were reported by McCammon (1970). The Harvard School of Public Health Growth Study was responsible for influential advances in methods for the radiographic measurement of subcutaneous adipose tissue and substantial findings relating to tissue growth and skeletal maturation.

The designs of the Fels Longitudinal Study and the Child Research Council Study were similar in many respects. Each was a study of individuals within families and, in each case, the examinations continued at intervals during adulthood. Physical growth and maturation dominated the Child Research Council program, but there was an approximate balance between research activities in physical areas and in psychological areas in the Fels Longitudinal Study. The balance altered in the Fels Study in 1974 when the collection of psychological data ceased and analysis of the existing psychological data slowed considerably.

### A little history
Until 1977, the Fels Longitudinal Study was conducted within the Fels Research Institute and it included concomitant studies of cognitive development, behavior, and family functioning. These studies have been excluded from the present account, which is restricted to physical growth, physical maturation and body composition. In 1977, the Fels Research Institute became part of the School of Medicine at Wright State University and the Fels Longitudinal Study was subsumed as the Division of Human Biology, at first within the Department of Pediatrics and later within the Department of Community Health. Many other investigations by scientists in the Fels Research Institute or in its successor have been omitted from the present account because they were not derived from the Fels Longitudinal Study, even if they relate to the themes of this book.

Many are puzzled by the location of the Fels Longitudinal Study and the Fels Research Institute in Yellow Springs, unaffected by the flow of resources and expertise to large well-known universities such as Harvard and Stanford. The answer concerns two close friends, Arthur E. Morgan and Samuel S. Fels, who were responsible for the foundation of the study and of the institute. In 1929, Arthur Morgan was the President of Antioch College in Yellow Springs. Therefore, it was natural that, from their conception to their maturity, the study and the institute that these men initiated were in Yellow Springs and associated with Antioch College.

Arthur Morgan posed a question that sounds simple: 'What makes people different?' This is one of many questions that sound simple but are extremely difficult to answer. Mr Morgan concluded that a longitudinal study from conception to adulthood was required. He discussed this with many biologists and social scientists but they were skeptical. Mr Morgan was not deterred; he approached Mr Fels, a Philadelphia businessman and philanthropist, about his plans and did not conceal the doubts of the scientists whom he had consulted. Mr Fels was not troubled by the opinion of the 'experts.' Although he realized the results of the study would be delayed, he gave it enthusiastic support and financial backing.

The parents of Samuel Simeon Fels had come from Bavaria to Philadelphia in 1848. They stayed there briefly before moving to Virginia where Samuel Fels was born in 1860. In 1873, his family returned to Philadelphia and achieved financial success in soap manufacturing. For Samuel Fels, financial success was not an end in itself but a means to do good. He used his wealth to assist the establishment and the work of many local, national and international causes that were related to education, labor laws, crime prevention, local government, science, and Jewish charities.

The early support by Samuel Fels of the program suggested by Arthur Morgan was given structure in 1932 when a corporation was established in Ohio. The purpose clause stated: 'The principal initial problem to which the organization shall devote its energies is the furtherance of knowledge of the effect of physical, emotional, and nutritional environment during and shortly after the period of gestation upon the physical and mental constitution of the child.' This is remarkable because, at that time, there was widespread scientific belief that a fetus was shielded from environmental effects. Time has shown the wisdom of the stated purpose which matched the attitude expressed by Samuel Fels in his 1933 book 'This Changing World.' He wrote that long-term studies of large groups of children should be made and they 'should begin before the child is born and continue to maturity.' In 1935, the Ohio Corporation was replaced by the Samuel S. Fels Fund which was incorporated in Pennsylvania with much broader charter purposes that matched the wide-ranging interests of Samuel Fels.

The Fels Longitudinal Study began in 1929 with the appointment of Lester W. Sontag as the Director. At that time, Lester Sontag was the Physician at Antioch College in Yellow Springs, having recently received his MD degree from the University of Minnesota. Dr Sontag received further training at various major centers in the US while he planned the Fels Longitudinal Study. The examinations of participants began in 1930 when the institute had a staff of three and an annual budget of $5000. While the institute was small, Lester Sontag was deeply involved in all the research and also managed administrative matters. Until his retirement in 1970, he continued to be active in research concerning skeletal maturation, skeletal variations and developmental changes in intelligence. His major contribution, however, was the foundation and nurturing of the Fels Longitudinal Study which he directed during a long period of growth and increasing effectiveness, and from its initiation under the protective umbrella of Antioch College to the status of a free-standing institution in 1947.

The Fels Research Institute was owned and operated by the Fels Fund of Philadelphia, with considerable assistance from the US Government through grants and contracts. This arrangement continued until 1977 when, as a result of increasing financial strains, the Fels Fund donated the institute to the School of Medicine at Wright State University, with an

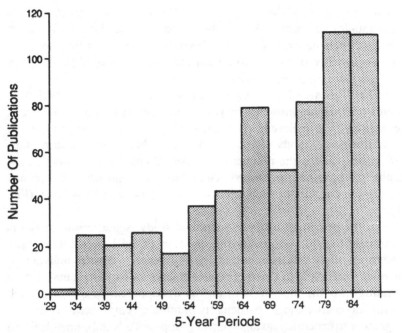

Fig. 1.1 The number of publications (articles, chapters, books, and monographs) dealing with physical growth and maturation from the Fels Longitudinal Study within 5-year periods. The numbers on the horizontal axis refer to calendar years at the beginning of these periods (e.g., '29 = 1929–1933).

agreement to fund the institute for a further 5 years. Wright State University is situated in Dayton, about 8 miles (13 km) from the institute. It was expected that the institute would be beneficial to the school of medicine that had just been established at the university by providing an immediate research base. This expectation was realized to some extent, but the institute was fragmented when part was moved to the main campus and it rapidly became weaker and smaller as scientists who resigned or retired were not replaced. Now, the institute is reduced to the Division of Developmental Psychology in the Department of Psychiatry and the Division of Human Biology in the Department of Community Health. Both these divisions are located in Yellow Springs near the building that housed the institute from 1947 through 1984.

On balance, the decision to locate the Fels Longitudinal Study and the Fels Research Institute in Yellow Springs was a good one. The local population is more stable in place of residence than is usual in the US; this stability is very important for a long-term serial study. Since the institute was not on a university campus, the professional staff were not burdened by teaching responsibilities and committee work. Nevertheless, some disadvantages accompanied the relative isolation of the Fels Research Institute. Each department in the institute lacked a critical mass of scientists and it was difficult to obtain the effective help of specialists in other disciplines, e.g., biostatistics.

The Fels Research Institute grew from its commencement in 1929 until about 1968, whether growth is judged by the number of staff or the size of the building. The achievements of the staff in research related to growth, maturation, and body composition are difficult to measure. One inadequate guide is the number of articles, chapters, books and monographs that have been published. The totals for 5-year periods remained fairly stable until about 1954, after which they increased markedly (Fig. 1.1).

From the beginning of the longitudinal study, some serious attempts were made to address directly the question posed by Arthur Morgan concerning what was responsible for differences between individuals. Analyses were made to determine the influence of environmental factors operating near the time of birth on the development of children. Such investigations are difficult, particularly in human beings, because random assignment to experimental and control groups is likely to be unethical and unacceptable. These topics were studied at Fels in what are sometimes called 'natural experiments' without controlling the conditions. With this approach, the data must be adjusted statistically for the influences of intervening variables. Consequently, large samples are required and data must be collected for many variables that could influence the results. Even then, the findings from such studies commonly remain inferential. Any observed effects may be due to another variable or set of variables for which data were not collected.

Instead of beginning with the recognition and measurement of effects, the institute staff, quite properly, began by describing the status and patterns of change with age in variables related to growth, maturity and body composition. In this they followed the wise advice of William Harvey: 'I am of the opinion that our first duty is to inquire whether a thing be or not, before asking wherefore it is.' Throughout its history, the Fels Longitudinal Study has resulted in many hypotheses that have been tested by the analysis of facts.

Until about 1945, almost all the research activities of the Fels Research Institute concerned the longitudinal study. In those days, the longitudinal study *was* the institute. In subsequent years, many other projects were located in the institute; some of these did not have direct relationships to the hypotheses that could be tested using data from the longitudinal study.

**Biographical sketches**

Brief biographical sketches follow, ordered alphabetically, of some of the scientists who worked at the Fels Research Institute and were leaders in the analysis of data relating to growth, maturation, and body composition. The list does not include the many scientists who worked at the Fels Research Institute in areas such as biochemistry, endocrinology, psychology and physiology. Others who came to Fels for short periods to work in the areas of growth, maturation and body composition so that they could share their special skills or acquire new ones are not listed. Many of these visitors and fellows reported analyses of Fels data, as is clear from the author index which contains the names of many distinguished scientitsts. The years in parentheses after each name in the following sketches give the period during which the scientist worked at the institute.

*Richard N. Baumgartner* (1985–1990). Rick Baumgartner received a PhD in Public Health from the University of Texas at Houston in 1982. He has contributed greatly to the longitudinal study by supervising the measurement procedures and by analyzing serial changes in patterns of adipose tissue distribution (fat patterns) and their relationships to risk factors for cardiovascular disease and to diabetes. In addition, he investigated new methods for the measurement of body composition including electrical impedance, computed tomography, and magnetic resonance imaging.

*Pamela Byard* (1981–1982). Pam Byard obtained a PhD degree in Biological Anthropology from the University of Kansas in 1981. While at the Fels Research Institute, she worked partly in Roger Siervogel's study of blood pressures within large family groups. She has also analyzed data from the Fels Longitudinal Study both while at Fels and after she left. These analyses concerned familial associations for growth, body composition and craniofacial variables and secular changes in body composition variables.

*Wm Cameron Chumlea* (1978– ). Cameron Chumlea obtained a PhD in Physical Anthropology from the University of Texas at Austin in 1978. He was responsible for all measurement procedures in the Fels Longitudinal Study until 1985 and plays an important part in the general design of the study. One of his major research interests is the assessment of nutritional status from body measurements, especially in the elderly. He is prominent in research related to adipocytes (fat cells), body composition, bioelectric impedance, the measurement of total body water, and the assessment of skeletal maturity.

*Christine E. Cronk* (1980–1982). Chris Cronk was awarded a DSc degree by the Harvard School of Public Health in 1980. While she worked in the Division of Human Biology, her research addressed the application of recently developed statistical methods to analyses of serial body composition data. She also developed national reference data and reported studies of growth and body composition in children with Down syndrome.

*Frank Falkner* (1971–1979). Frank Falkner received his medical training in England where he graduated in 1945, becoming a Fellow of the Royal College of Physicians in 1972. He succeeded Lester Sontag as the Director and steered the institute through a period of financial and administrative difficulties. In addition, he established the Fels Division of Pediatric Research at the University of Cincinnati and began a multinational study of infant mortality. While at Fels, Dr Falkner's personal research related to the assessment of nutritional status, infant mortality, and the joint editorship, with James M. Tanner, of a three-volume treatise on human growth.

*Stanley M. Garn* (1952–1968). Stanley Garn obtained a PhD in Physical Anthropology from Harvard University in 1948. During a stay of exactly 16 years as Chairman of the Department of Growth and Genetics, he gave the institute national and international prominence by his many publications and his answers to novel questions. Dr Garn's major research contributions related to skeletal and dental maturation, age changes in subcutaneous adipose tissue, the gain and loss of bone substance, and the sizes and shapes of teeth.

*Shumei Guo* (1985–   ). Shumei Guo received her PhD in Biostatistics from the University of Pittsburgh in 1983. She is in charge of all statistical matters in the Division of Human Biology, including the management of our large and complex data bases. In addition to acting as a statistical consultant to other staff members, Dr Guo is particularly active in the development of new methods for the analysis of serial data, in the use of new procedures to construct equations that predict body composition variables from values that are relatively easy to obtain, and in the analysis of risk factors for cardiovascular disease.

*John H. Himes* (1976–1979). John Himes obtained a PhD in Physical

Anthropology from the University of Texas at Austin in 1975. While working in the Division of Human Biology, he supervised data collection for studies of body composition and hearing ability. His main research interests concerned bone lengths, the estimation of stature from bone lengths, the correct interpretation of observed statures, the hearing ability of children in relation to noise exposure, the assessment of nutritional status, body composition, and the provision of reference data for growth rates.

*Harry Israel III* (1964–1975). Harry Israel obtained his DDS degree from the University of Michigan in 1956 and a PhD degree from the University of Alabama in Birmingham in 1971. During the long period that he worked part-time at the Fels Research Institute, his research interests were related to growth and aging in the craniofacial part of the skeleton, variations in the bones of the fingers, aging of vertebrae, and the radiographic density of bones and teeth.

*Arthur B. Lewis* (1939–    ). Arthur Lewis obtained his DDS degree from Ohio State University in 1933 and was awarded an MS in Orthodontics by the University of Illinois in 1935. He has a remarkable half-century record of part-time voluntary work at the Fels Research Institute. His research publications describe the sizes and shapes of teeth, dental development including its genetic control, the occasional failure of teeth to develop, and craniofacial growth.

*Debabrata Mukherjee* (1980–1984). Debu Mukherjee received a PhD in Biostatistics from the University of Alabama in Birmingham in 1980. At the Fels Research Institute, he was responsible for statistical analyses and the management of data bases. His main publications concerned serial analyses related to body composition, hearing ability, craniofacial growth, the development of equations to predict body composition variables, infant growth patterns, and the measurement of skeletal mass.

*Earle L. Reynolds* (1943–1951). Soon after he obtained a PhD degree from the University of Wisconsin, Earle Reynolds came to the Fels Research Institute as Chairman of the Department of Growth and Genetics. While at Fels, he published landmark papers dealing with age changes in adipose and other tissues, the growth of the pelvis, and methods for grading the development of secondary sex characters.

*Meinhard Robinow* (1939–1943). Meinhard Robinow obtained his MD degree from the University of Hamburg in 1934. While at the Fels Research Institute, he made very thorough physical examinations of the participants and recorded his findings in a script that was very easy to read. His research publications related to the maturation of bones and teeth, methods for grading the posture of children, and the provision of reference data for growth rates.

*Alex F. Roche* (1968–   ). The present author graduated in medicine from the University of Melbourne (MB, BS) in 1946 and subsequently received PhD, DSc and MD degrees from the same university before becoming a Fellow of the Royal Australasian College of Physicians. He came to the institute as Chairman of the Department of Growth and Genetics and was later Head of the Division of Human Biology and Director of the Fels Longitudinal Study. He is in charge of research relating to physical growth, maturation, and genetics. His research interests include growth of the whole body and of the craniofacial region, skeletal maturation, the prediction of adult stature, body composition, risk factors for cardio-vascular disease, hearing ability in relation to noise exposure, and methods for the analysis of serial data.

*Roger M. Siervogel* (1974–   ). Roger Siervogel obtained a PhD in Human Genetics from the University of Oregon in 1971. He has led the Fels Division of Human Biology into the computer age and is active in research related to body composition, risk factors for cardiovascular disease especially blood pressure, dermatoglyphic patterns, hearing ability in children in relation to noise exposure, and genetic and familial factors that affect these traits.

*Lester W. Sontag* (1929–1970). Soon after graduating in medicine from the University of Minnesota in 1926, Lester Sontag became the resident physician at Antioch College, and in 1929 he was appointed as the Founding Director of the Fels Research Institute. His research work, in relation to the topics covered in this book, concerned prenatal growth, nutrition, genetics, the growth and maturation of bones, and basal metabolic rate. After 1945, Dr Sontag's interests became more behavioral and his administrative activities occupied more of his time.

### The major US longitudinal studies
Sontag (1971)[281], in an illuminating review of the major longitudinal studies in the US, drew attention to the changes that had occurred in these studies since they began in the late 1920s and early 1930s. All of them

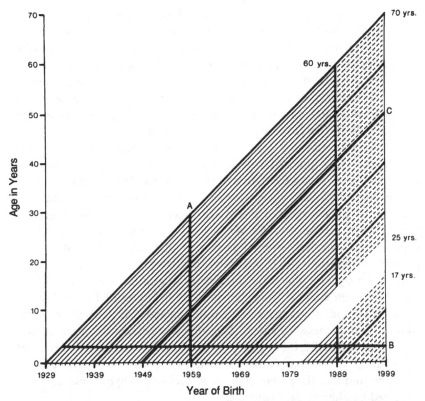

Fig. 1.2 The design of the Fels Longitudinal Study with lines drawn to illustrate subsets of the data that could be analyzed separately. An analysis of data crossed by line A could furnish information about the status of participants in 1959. The data crossed by line B could provide data about 2-year-old children, and those exemplified by line C could be informative about patterns of growth.

started with a global approach – the aim was to study the whole child until adulthood with an emphasis on the variations between children in their patterns of change. This serial multidisciplinary goal was a tall order for which the measurement and analytic procedures were grossly inadequate at that time.

As a result, masses of data were collected in the major US longitudinal studies. Many of the variables were of dubious value in the study of normal children although data of the same type were useful in clinical diagnosis. The studies began without hypotheses and this, in large part, explains the unselective accumulation of data. Almost all the early reports were cross-sectional and descriptive. The decision to publish cross-sectional reports was mandated by the nature of the data base early in the studies but it continued long after analyses of serial data became possible. Cross-sectional analyses have considerable appeal because it is easy to present the findings from such analyses in ways that are clinically useful and easily understood, unlike findings from complex serial analyses. Due to the lack of balance between data collection and data analysis and the unimaginative nature of the analyses, the more productive scientists tended to leave the longitudinal studies during the 1930s and 1940s, while those who were satisfied with the accurate collection of data tended to stay.

This situation did not last. The studies were forced to change as the foundations that supported them became discouraged at the small number of research reports and the nature of these reports. The changing attitude was expressed rather vaguely by Reynolds (1949a) who wrote: '... longitudinal data appear to give much superior information on growth trends.' More precise attitudes were needed. The studies began to address hypotheses that centered around the concept that the conditions that affect growth, maturation or risk factors for disease may precede these phenomena by long intervals. This change in attitude was accelerated in the 1950s when support from the National Institutes of Health became available through a grant mechanism that included reviews by scientists who were hypothesis oriented. This did not necessarily mean a new beginning. Gradually, the scientists within the studies realized that some of the existing serial data could be used to answer important research questions and that the potential of the studies to test hypotheses could be increased by altering the protocols.

### The Fels Longitudinal Study
The design of the Fels Longitudinal Study matched that of the Child Research Council Study that had started in Denver, Colorado, two years before the former began. Originally the Fels Study was to continue until the participants were 16 years old, then its termination was delayed to 18 years, and still later it was extended into adulthood without mention of a final age. These extensions have added to our ability to answer research questions of great health-related significance.

The design of the Fels Longitudinal Study is shown in Fig. 1.2. The horizontal axis shows the calendar years since 1929 when the study commenced and the vertical axis refers to the ages of the participants. The oblique lines trace the passage of each annual cohort as its members become older. The vertical line passing through 1989 represents the recent state of the study as it cuts across the lines for cohorts at ages from birth to 60 years. The interrupted oblique lines to the right of this vertical line represent the planned enrollment of additional cohorts and the changes

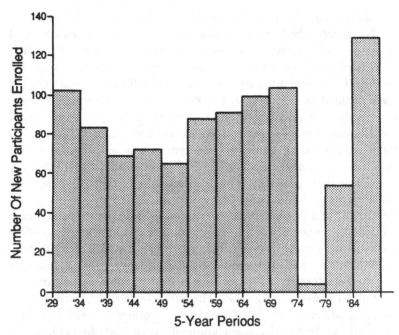

Fig. 1.3 The number of new participants enrolled by 5-year calendar periods; the numbers on the horizontal axis refer to the calendar years at the beginning of each period (e.g., ′29 = 1929–1933).

that will occur in the ages of all participants up to 1999. Figure 1.2 does not indicate one important change with time in the protocol for the Fels Longitudinal Study. Since it began there has been a marked increase in the number of variables measured; the increase has been very rapid since 1974 and became possible at this time when the collection of psychological data stopped. The present physical program, in terms of the number of procedures and the time required for a set of examinations, would have been too onerous for the participants if combined with the previous psychological program.

This diagram of the Fels design in Fig. 1.2 represents not only the pattern of enrollments and examinations but also the nature of the data base; it can be used to illustrate some of the analytic possibilities. A vertical slice through the data would allow, for example, the analysis of all the measurements recorded in 1959 as indicated by the vertical line (A) in Fig. 1.2. The results could provide information about the status of individuals between birth and 30 years of age. Alternatively, a horizontal slice (B) could be made through the data as shown by the line drawn through the figure at the 2-year level. The analysis of the data in this slice could provide information about 2-year-old children born in different years; conclusions might be reached about possible secular changes and the possibility of combining (pooling) data from neighboring cohorts for an analysis. The third possibility is to cut an oblique slice through the data set as indicated by the line (C) in the figure that begins at 1949 on the horizontal axis. The analysis of the data in this oblique slice could provide information about changes within individuals. Such an analysis is not likely to be effective if restricted to those enrolled in one calendar year because the annual cohorts are small. If cohorts are combined, this type of analysis presents exciting possibilities that cannot be achieved without access to longitudinal data. This third analytic possibility directs attention to the exciting and very unusual nature of the Fels data base.

Between 12 and 20 participants were enrolled into the Fels study each year during the pregnancy of the mother, and the first measurements of the participants were made near the time of birth. At the time of enrollment, all the families lived in the southwestern part of Ohio within 30 miles (48 km) of the Fels Research Institute. Now, five of the active participants live abroad and 150 live outside Ohio. The numbers enrolled, during 5-year periods, are shown in Fig. 1.3. The rate of enrollment was relatively high early in the study but decreased gradually until 1949 to 1953. It has increased since then, except for the period 1974 to 1981 when enrollment was suspended due to lack of funds.

There are 1036 Fels participants, of whom 537 have been examined during the past 8 years. In addition, data have been recorded from many non-participant relatives who have been examined less frequently. Some participants were lost from the study due to early death (miscarriages, stillbirths, neonatal deaths), but most of the loss of participants from the study reflects unwillingness to continue and changes in places of residence to unknown locations. The loss (attrition) from the study has been random in regard to some variables, e.g., weight, at the most recent examinations.

All the participants are white except for 15 who are black. Blacks are underrepresented in the study by any measure; they form about 25% of the Yellow Springs population. Blacks were welcome in the study but few volunteered. Plans have been developed recently to enroll a large number of blacks. At enrollment, about 35% of the families lived in cities of medium size (population 30 000 to 60 000), about half in small towns (population 500 to 5000) and the remainder lived on farms. The families of the participants have a wide range of socioeconomic status (Sontag, Baker & Nelson, 1958[283]; Crandall, 1972 and unpublished) as shown by the distributions of scores on the Hollingshead Two Factor Index of Social Position (1957). The distributions of these scores for the Fels families are similar to those in national US samples except for an underrepresentation of the lowest of the five Hollingshead groups for the families of participants born after 1939.

When the participants were 6 years old, about 23% of the fathers were major professionals, 16% were lesser professionals, executives or business-men, 26% were minor professionals, executives or businessmen, 9% were clerical workers, technicians or salesmen, 16% were skilled manual employees, 6% were semi-skilled manual employees, and 2% were unskilled or unemployed. Bachelors degrees had been obtained by 39% of the fathers and 27% of the mothers. Masters or doctoral degrees had been obtained by 13% of the fathers and 3% of the mothers.

The Fels Longitudinal Study is based on a sample of convenience. Participants were enrolled at the request of pregnant women from local families considered to be 'stable.' While 'stable' was not defined clearly, it included judgments of geographic stability and of attitudes likely to be associated with a long-term commitment.

There are four sets of triplets and 14 sets of twins, including five sets of identical twins, amongst the Fels participants. In all analyses, only one random member of each identical pair has been included and triplets have been excluded from some analyses because they are overrepresented in the sample. The non-identical twins have been included in most analyses because their growth data do not differ significantly from those of the other participants. Gestational age is poorly established for many of the participants but birth weight is known. At birth, 6.6% weighed less than 2500 g, 2.1% weighed less than 2000 g and only 0.4% weighed less than 1500 g. These frequencies are in reasonable agreement with the US national distribution of birth weights.

Fourteen of the participants were abnormal at birth and 36 developed serious chronic illnesses postnatally. These participants have followed the study protocol but their data have been excluded from analyses if there was any possibility that the pathological condition could have influenced the variables being analyzed.

After the Fels Longitudinal Study had been conducted for a few years, some of the mothers became pregnant again and requested that their next children be enrolled. These requests were granted and siblings were enrolled. Totally, 344 offspring of participants (second generation), and 90 of their offspring (third generation) have been enrolled. Some terms used within the Fels study need clarification. 'First generation participants' are those who do not have a parent enrolled in the study since birth. Those with a parent who has participated in the study since birth are called 'second generation participants' and those with a grandparent and a

Table 1.1. *Tolerances* ($\pm$ *days*) *for ages at examinations in relation to scheduled ages*

| Scheduled ages (years) | Tolerances | Scheduled ages (years) | Tolerances |
|---|---|---|---|
| 6 months and younger | 2 | 4.0 | 14 |
| 9 months | 3 | 4.5 | 16 |
| 1 | 4 | 5.0 | 17 |
| 1.5 | 5 | 5.5 | 18 |
| 2.0 | 7 | 6.0 | 22 |
| 2.5 | 9 | 6.5 | 23 |
| 3.0 | 11 | 7.0 | 25 |
| 3.5 | 13 | 7.5 and older | 30 |

parent who have participated in the study since birth are called 'third generation participants.' There is now one fourth-generation participant – a girl whose great-grandparent has been studied since birth.

The active participants can be grouped into about 200 kindreds, some of which contain several nuclear families. There are from one to seven participant siblings within each nuclear family. In addition, data have been obtained from unenrolled spouses when they were adult and from unenrolled siblings at 7, 11 and 17 years of age.

Advantages and disadvantages are associated with the familial relationships among the Fels participants. The advantages relate to the ability to analyze associations between family members for measured variables. This can be done in the Fels study using data collected at matching ages, for example, in pairs of mothers and daughters both measured at 12 years; this is not possible with any other existing data set. The disadvantages concern potential statistical bias if related individuals are included in some types of analyses. Only small problems are caused by this type of bias in the Fels study.

The examinations were scheduled at fixed chronological ages that were not related to the presence of illnesses (mothers were asked to bring their children for examination whether well or sick) or the achievement of developmental landmarks. Through 1973, the scheduled ages were 1, 3, 6, 9 and 12 months, then each half-year to 18 years in each sex. After 1973, the half-year examinations were omitted from 5 through 10 years and from 16 through 18 years in the boys and from 5 through 9 years and from 14 through 18 years in the girls. The age ranges for these omissions are periods when the rates of growth differ little with age. After 18 years, the examinations were scheduled at 20, 22 and 24 years. Examinations after 24

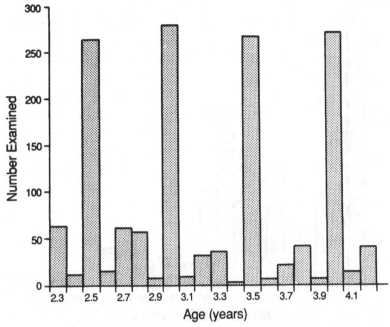

Fig. 1.4 The frequency of the actual ages at examinations for girls from 2.3 through 4.2 years; note the occurrence of some examinations at each 0.1 year interval despite the efforts to measure at the scheduled ages (2.5, 3.0, 3.5, 4.0 years).

years have been less regular. They have been dependent on extramural funding since 1976 when all examinations after 8 years were subsumed into the Body Composition Study which is described in Chapter 7.

Age tolerances were set for the examinations from the beginning of the Fels study (Table 1.1) because it was realized that variations in ages at examinations would influence the results. For example, the mean increase in stature from 4 to 5 years can be determined accurately as the difference between the mean at 4 years and the mean at 5 years, if all the participants were examined at exactly 4 years and 5 years. If, however, the ages at examinations vary, the difference between the pair of means labeled as '4 years' and '5 years' will only approximate the true mean increase from 4 years to 5 years.

Despite all efforts, many participants were measured at ages outside the tolerances. An example of the distribution of actual ages at examinations is shown in Fig. 1.4. This figure shows the number of examinations of girls at each 0.1 year interval from 2.3 to 4.2 years. Despite the large peaks at 2.5, 3.0, 3.5 and 4.0 years, a considerable number of examinations was

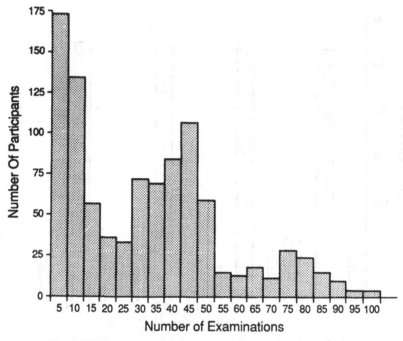

Fig. 1.5 The number of participants with particular numbers of examinations (5 = 0 to 5 examinations; 10 = 6 to 10 examinations, etc.). The large number with fewer than 10 examinations reflects continuing enrollment at birth.

made at intermediate ages. Until 1972, it was not common to use data recorded at examinations outside the age tolerances in analyses because it was considered they were not useful. Later, it was realized that estimated values for scheduled ages could be made from the serial data. A detailed description of these adjustments to the data is given in Chapter 2.

There are large variations among the participants in the number of times they have been examined (Fig. 1.5). These variations reflect the continuing enrollment into the study. Many of those with few examinations are still very young. A total of 720 participants has been examined more than 10 times and 440 have been examined more than 20 times. The total number of examinations for the group exceeds 24 000. The number of participants with records of particular lengths is shown in Fig. 1.6; about 370 have records extending more than 25 years.

The data recorded or measured in the Fels study may be grouped as follows.

> *Physical growth*: total body size, body proportions, length and sizes of segments.

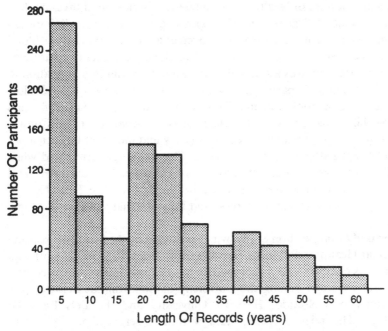

Fig. 1.6 The number of participants with records of particular lengths (5 = 0 to 5 years; 10 = 6 to 10 years, etc.).

*Physical maturity*: body photographs and ages at reaching developmental milestones, grades of secondary sex characters, menarche and menopause, skeletal and dental ages, and ages at peak height velocity and peak weight velocity.

*Skeletal and dental data*: sizes and shapes of bones and teeth, skeletal mass, skeletal variations.

*Body composition*: total body composition (total body fat, fat-free mass and percent body fat from underwater weighing, impedance and total body water, dual photon absorptiometry), regional body composition (thicknesses and areas of subcutaneous adipose tissue, muscle and bone), grip strength, and adipocyte size.

*Other*: risk factors for cardiovascular disease (blood pressures, lipids and lipoproteins), dietary intakes, genetic markers, data from physical examinations and medical histories, hearing ability, noise exposure, menstrual histories, and histories of physical activity.

There have been some extensions to the basic plan of the Fels study. The first major extension occurred in 1972 when a contract was awarded by the

National Institutes of Health for the analysis of serial blood pressure data from the major US growth studies. This began a spirit of collaboration among these studies and led to improvements in the recording of blood pressures at Fels and in the methods by which we store data in computers. The next extension was a study of hearing ability in children in relation to noise exposure. Fels participants were included along with some Yellow Springs High School students. This serial study, which lasted 8 years (1975–1982), was funded by the Office of Noise Abatement and Control within the Environmental Protection Agency and monitored by scientists from Wright-Patterson Air Force Base. The study was terminated when the Office of Noise Abatement and Control was closed because the Federal Government decided noise was not a health hazard. The Fels Study of Hearing Ability remains the largest and longest serial study in this topic area.

The Body Composition Study, which is funded by the National Institute for Child Health and Human Development, began in 1976. The original design was complex and its complexity has increased. It commenced as a collaborative study among the University of California at Berkeley, the University of Colorado, Harvard University and the Fels Research Institute. The relevant serial data were transferred from the other institutions to Fels, follow-up studies were made at the University of California and at Harvard University, and an intense program to record body composition variables began at Fels. The Fels Body Composition Study, which in its present cycle is funded through 1993, is the largest serial body composition study anywhere. Many of the data recorded in this study are relevant to the risk of cardiovascular disease in middle age, for example serial measurements of blood pressure and blood lipids. It is one of three major long-term US studies of serial changes in cardiovascular risk factors during childhood and adulthood.

### The well-planned long-term longitudinal study

Few have the opportunity to work in longitudinal studies for as long as the present author, who began the University of Melbourne Growth Study in 1954. Therefore, it is appropriate to weave into this text some reflections on the design of such studies.

There are many reasons why it is difficult to plan a longitudinal study of human growth; the implementation of such a study may be even more difficult. The study must continue for a long period but not necessarily for the whole life span. To take a simple example, a longitudinal study of premature infants extending to 6 years could test many hypotheses. An adequate design for such a study could be based on a single cohort of all premature infants born in a selected county during a specified interval. Some infants would be omitted, however, because their mothers chose not to participate and there would be attrition from the sample during the study. It is unlikely that the measurement procedures would become outmoded during the 6-year study, but the applicability of the findings will be reduced if there are major changes in the management of premature infants either during or after the study.

A series of such studies could be planned that, in combination, would cover all ages from birth to adulthood. Ages at enrollment would vary: perhaps birth for one cohort, 6 years for another, etc. This approach has considerable appeal because, in comparison with a single study based on one cohort, the work will be completed quickly with less attrition and less risk that the hypotheses will cease to be of interest. A group of such studies, conducted simultaneously, should be associated with greater investigator interest but will probably require a large staff and substantial funding.

There are problems in designing a serial study that may extend from birth to adulthood or continue throughout the whole life span. Such a study should be contemplated only if many important hypotheses are to be tested that require very long-term serial data. Such hypotheses could relate status or change during childhood to adult status. Some have claimed that human beings cannot conduct such studies because they will not live long enough (Falkner, 1973b), but such studies are possible with a succession of investigators (Roche *et al.*, 1981a[235]).

The study should be multidisciplinary only if the hypotheses require it and there is a convincing rationale. The multidisciplinary aspects must be included during the design phase (Roche, 1974b[179]). The mere existence of serial data for several disciplines within a single study does not lead to multidisciplinary analyses. These analyses require, as stated above, appropriate hypotheses and also real interest by investigators who work in different disciplines but have mutual respect.

The study design should include a plan to test sequential groups of hypotheses; some will require data from infants, others will require data from children or adults. It is unlikely that all the final hypotheses will be formulated before the study begins, but a comprehensive listing should be attempted and revised as the study proceeds. This list should be made as a template for progress, not just as a mechanism to obtain funds that will be forgotten as soon as possible.

As stated earlier, the design should be dependent on analytic considerations and, of course, these will depend on the hypotheses. It can be assumed that the serial nature of the data will be taken into account in the analyses, but the exact serial nature of the data, and thus the design, will be determined by the specific hypotheses and the analytic approaches they require. Is this a follow-up study for which the analytic procedure may be logistic regression which requires data for a pair of ages? Is it a study of patterns of change within individuals requiring a substantial number of serial data points for each individual to which a mathematical function will be fitted? If the latter, there must be short intervals between examinations, particularly when growth is rapid or there are large changes in growth rates. If the hypotheses are similar in nature for the different measurements, the same set of intervals between examinations should be suitable for most body measurements, except those of the craniofacial region for which a relatively large proportion of growth occurs during infancy. A different set of intervals may be suitable for other variables such as blood lipids.

The specific hypotheses and the analytic procedures to be applied will affect the sample sizes that are required. The sample sizes needed are likely to differ among hypotheses. If some of the hypotheses relate to groups near the extremes of the distributions, large samples or specially selected samples should be enrolled. Those planning the study must decide the level beyond which a further increase in sample size to allow testing additional hypotheses is not justified.

The examinations must be scheduled at predetermined target ages that are independent of possible illness. Despite all efforts, many ages at examination will differ from the target ages (see Fig. 1.4). Until about 1970, most of those who analyzed serial data excluded data gathered outside the tolerances for target ages. It is now realized that all the recorded data are informative and that none should be disregarded.

A long-term study of human growth is very expensive and it is unrealistic to expect a funding commitment for a period longer than about 5 years. Therefore, the hypotheses and the design must facilitate the publication of research reports based on tests of hypotheses during the first few years of the study at a rate that will convince the funding agency to continue its support and will encourage the continued participation of key scientists. These considerations affect the enrollment plan. A choice could be made between one large cohort of newly born infants, or small annual cohorts. If one large cohort is enrolled, some analyses can be made early. This is not the case if small annual cohorts are enrolled.

Enrollment of a large cohort at birth is associated with logistic problems due to the need for closely spaced examinations during infancy. The work load may be overwhelming at a time when funding is restricted because the funding agency lacks confidence. In the early phase of a study, recording forms must be prepared, informed consent obtained, data entry procedures established, and computer programs written. At the same time, the staff must be trained and monitored in data collection procedures. There is a risk that the single large cohort may be unrepresentative or that one or more important variables may be omitted at the beginning of the study. There is no practical way to remedy such defects. If all enrollments were at birth, the hypotheses could be tested in an order determined by the sample size required for this testing. If, however, the design includes the enrollment of annual cohorts, their representative natures can be compared and variables omitted from the initial design may be added for later cohorts. On the contrary, variables that prove unrewarding can be discontinued. In some situations, an intermediate design may be appropriate in which cohorts of medium size are enrolled at, say, 5-year intervals.

Another design that can be considered is the enrollment of a single cohort of families each of which includes a newly born infant. The older participants within these families will require less frequent examinations than the younger participants and hypotheses that require only short-term data can be tested early in the study for numerous age groups. In addition, associations within families can be analyzed. The introduction of bias is a disadvantage but this can be overcome by including only one randomly chosen participant of each sex from each family in a specific analysis, when the need for this is demonstrated. An additional disadvantage of this design is the delay that would occur before sufficient data were available to allow analyses relating to particular age groups.

There is no perfect solution; planning is never perfect. With any enrollment plan, it is unlikely that a random sample will be obtained for a longitudinal study. Therefore, generalizations to a total population may not be defensible. If the effects of specific environmental variables are to be estimated, in the absence of random assignment to experimental and control groups, many intervening variables must be recorded so that statistical adjustments can be made. The need for these adjustments increases the sample sizes required to test the hypotheses. Measuring a large number of variables is expensive and requires considerable staff time for data collection and data entry. It also increases the burden on the participants. Consequently, as the study proceeds, variables should be added only after careful attention has been given to the possible consequences. Finally, it is essential to protect the study participants, who will already be burdened by the planned examinations, from other scientists who may regard them as a ready pool of volunteers for additional projects.

Having enrolled the participants, great efforts must be made to ensure that they do not miss examinations, that they complete all the procedures scheduled for each examination, and, most importantly, that they do not withdraw from the study. Even if the examination procedures are painless and non-invasive, these goals will not be fully achieved. Various rewards and study newsletters help but the most important element is a set of staff attitudes and behaviors that make an examination a pleasant experience. On balance, the enrollment of families rather than individuals probably increases long-term participation in a study.

All measurements must be made in a standardized fashion that provides valid reliable data, as stressed by Roche (1974b[179]). Therefore, one early goal of a long-term study may be the validation of some measures. The study may be an excellent milieu for testing new procedures but this testing should involve separate groups of subjects recruited for this purpose. Reliability of data is an on-going concern that will continue throughout the study. Additionally there is always the possibility that repeated examinations may alter the phenomena being measured leading to confounding of age effects and examination effects. It is possible to separate these effects for a procedure that is added to the protocol in a chosen calendar year and is applied to participants of all ages (Roche *et al.*, 1983a[255]).

When there are improvements in equipment or techniques, they should be incorporated cautiously into the study protocol, in accordance with the view of Alexander Pope:

> Be not the first by whom the new are tried,
> Nor yet the last to lay the old aside.

As a long-term study proceeds, it is likely that changes will be needed in the variables recorded and, perhaps, the ages at examinations. These changes will be justified if they assist the testing of the original hypotheses or other hypotheses that are closely related to them. Additional hypotheses are justified only if they relate to the original topics. New topics should be added to the study only if the data collected earlier are relevant to them. The general principle is that all additions should strengthen or extend the original study, the link between the original and the added parts must be more than a common sample of participants.

Secular trends may be important in the interpretation of findings from a long-term study. These trends, that usually reflect decades of slow change, are, in a strict sense, alterations in recorded values with calendar year within a fixed population that does not gain or lose members by migration or differential mortality. Commonly, however, the term 'secular trends' is applied, somewhat loosely, to a national population despite migration and differential mortality. Generally, secular trends reflect alterations in the environment that may involve nutrition, illness, health care, pollutants, or other factors. Cross-sectional studies are commonly confounded by secular changes in the features measured and by selective mortality. In longitudinal studies, each participant acts as his or her own control and changes with time can be estimated more accurately. It may be desirable, however, to conduct parallel cross-sectional studies of larger groups so that secular trends in a more general population can be evaluated and used to interpret the findings from the longitudinal study.

Secular trends can be a nuisance to a data analyst because they make it necessary to analyze the data within subgroups based on birth years or the data may have to be adjusted for these trends prior to further analysis. The trends may, however, be of real interest and, if analyzed together with intervening variables, may direct attention to possible causes of the changes with year of birth in the variables studied. It is unlikely, however, that the cause of the changes will be established within a serial growth study; this requires a different design.

I wish to end this chapter with a note about collaboration. Long-term serial studies are rare and they are expensive. Therefore, it is important that the recorded data be utilized as fully as possible. Widespread realization of this is the basis for the cooperation that now exists among all the US longitudinal studies, and between them and the major cross-sectional studies in the same topic areas. Due to this spirit, and the availability of electronic data transfer, information from other studies is used commonly to replicate findings. This occurs frequently at Fels, and we are most grateful to our colleagues who make it possible.

# 2 The management and analysis of data

No human investigation can be termed true science if it is not capable of mathematical demonstration.

*Leonardo da Vinci* (1452–1519)

In the Fels Longitudinal Study, as should occur in any long-term serial study, great efforts were made to ensure that the data collected were reliable and that this reliability was retained during the transfer of the data to computers. High levels of data quality can be achieved in a prospective longitudinal study but not in a retrospective study. Additionally, the hypotheses posed and the analyses made in the Fels Longitudinal Study ensured, as far as possible, that the maximum information was derived from the serial nature of the data. Aspects of data management and analysis in the study will be described in the sequence: (i) the need for accurate data, (ii) quality control, (iii) data management, (iv) interpolation, (v) derivation of variables, (vi) transformation of variables, and (vii) statistical analyses.

### The need for accurate data

In a cross-sectional study, errors in the measurement of some individuals have little effect on the results of analyses unless these errors are large and common. If it is concluded that outlying values denote abnormal individuals or that large errors occurred during data collection, these data points can be excluded from cross-sectional analyses. This exclusion should be documented, and based on objective rules.

Large errors may be detected by comparing observed data with the distribution of values for the same variable in other groups of children of the same sex and age. For example, a recorded stature of 90 cm for a 6-year-old boy can be recognized as an outlier by comparison with the 5th

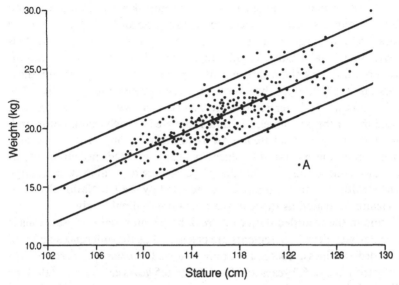

Fig. 2.1 A linear regression of weight on stature for 6-year-old boys in the Fels study. The upper and lower oblique lines ($\pm$ 2 SD) indicate the limits of the normal range for this relationship. Point A indicates an unusual combination of values that may be associated with an error of measurement or an abnormality.

percentile level (108.5 cm). An alternative is to use a bivariate regression that shows the relationship between two variables. The second variable should be one that has a constant or proportional relationship to the first. If weight is regressed on stature for 6-year-old boys in the Fels study, the combination of a stature at about the 95th percentile level of national data with a weight at about the 5th percentile level will be apparent as an outlying pair of values although the stature-for-age and the weight-for-age are independently acceptable (point A in Fig. 2.1).

An abnormal bivariate relationship of weight to stature for an individual could be due to a truly unusual value for weight or stature or to a recording error that affected either the value for weight or that for stature. Other data may be available, such as sitting height which is highly correlated with stature. The normality of sitting height relative to the recorded stature for the individual can be determined using a bivariate regression. If the recorded sitting height is within the expected range for the recorded stature, this provides presumptive evidence that the recorded stature is accurate and that suspicion should be directed at the value for weight.

Errors of measurement or recording have much more serious effects in longitudinal studies than in cross-sectional studies. Consequently, every effort must be made to avoid them. Some mistakes sound obvious and easily avoidable, such as the incorrect listing of sex or birth date, but these errors occur in large data sets. Fortunately, they are rare in the Fels Longitudinal Study. Other errors are common but almost all are minor. Errors occur during the measurement of participants but, generally, these errors are much smaller than those in large cross-sectional studies. The participants in the latter studies are not familiar with the measurement procedures or the personnel, less time is devoted to each examination and, commonly, the personnel are less aware of the need for care and accuracy. For example, the median differences between repeated measurements of stature for children aged 12 through 17 years were more than twice as large in the Health Examination Survey, conducted by the US National Center for Health Statistics, as those in the Fels Longitudinal Study.

Errors in the recorded data can wreak havoc on serial analyses. This is particularly evident if increments are calculated as the differences between the recorded values at successive examinations. For example, increments in stature from 5.5 to 6.0 years and from 6.0 to 6.5 years can be calculated for individuals by subtracting their statures at 5.5 years from their values at 6.0 years, and by subtracting their statures at 6.0 years from their values at 6.5 years. Any participants whose values at 6 years were recorded higher than reality will have erroneously high calculated increments for the interval 5.5 to 6.0 years and erroneously low calculated increments for the interval 6.0 to 6.5 years. Thus a single recording error leads to two errors in the calculated increments.

Reference data for increments are useful clinically partly because of their conceptual simplicity and because they treat growth as a dynamic process. Nevertheless, more complex mathematical approaches are needed to describe the patterns of growth of individuals. This need was recognized by Palmer and Reed as long ago as 1935. They wrote: 'If long series of observations are available it will be found advisable, probably, to derive individual growth curves and thus to make the final analysis of growth in terms of the parameters or mathematical characteristics of these curves.'

The description of growth data for an individual by a mathematical function (curve-fitting) is basic to the modern analysis of serial data. This procedure is very effective if an appropriate function is chosen, but there may be some individuals in the group for whom the function does not fit the data. This may occur because these individuals differ from the remainder of the group in their growth patterns. If this is the case, the divergence from the common growth pattern, shown as the difference between the recorded data and the fitted curve, will typically involve the data recorded at successive examinations and will involve more than one

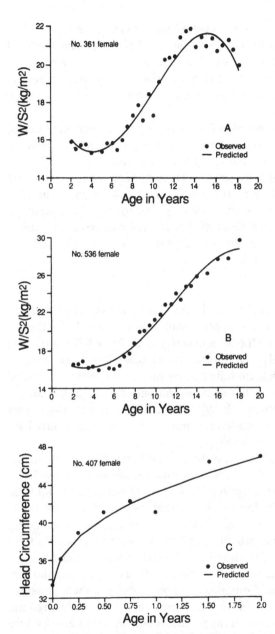

Fig. 2.2 Examples of serial data to illustrate how curve-fitting can identify variations in growth patterns (A, B) and erroneous points (the value at 1.0 years in C).

variable. The changes with age for weight/stature$^2$ (kg/m$^2$) in about 80% of Fels participants could be fitted by a family of functions that declined after about 14 years, as shown in Fig. 2.2A for participant number 361. In most of the remainder, there was no decrease after 14 years and, therefore, a different model was fitted, as shown for participant number 536 (Fig. 2.2B).

Curve-fitting is not only essential for the description of growth patterns, it directs attention to measurement errors. A large discrepancy between the recorded data and the fitted values may reflect measurement error. This is illustrated by the data for head circumference in participant number 407 (Fig. 2.2C). The recorded point at 1.0 years is markedly smaller than the value recorded at 0.75 years, which would lead one to suspect the accuracy of the points. With computer-based quality control such erroneous points would be recognized while the participant is still present and the measurement would be repeated.

### Quality control

Quality control has been emphasized in the Fels Longitudinal Study since its inception. The actions to ensure high quality data have included the use of equipment that is state-of-the-art and carefully calibrated. Records have been kept of the monthly calibration results since 1969. The anthropometric procedures, which are fully documented in volumes of *Standard Operating Procedures*, match the recommendations of the Airlie Consensus Conference (Lohman, Roche & Martorell, 1988) except that chest circumference, mid-thigh circumference and the anterior thigh skinfold are measured at slightly different levels.

In the Fels Longitudinal Study, chest circumference is measured at the level of the nipple, except in women for whom it is measured just inferior to the breasts. The Airlie Consensus Conference recommended that it be measured at the level of the fourth costosternal junction. Identification of the latter level would be too invasive for a longitudinal study including women. The mid-thigh circumference and the anterior thigh skinfold are measured at the junction of the middle and distal thirds of a line from the anterior superior iliac spine to the proximal border of the patella in the Fels study. The Airlie Consensus Conference recommended that these measurements be made at the mid-point of a line from the center of the inguinal ligament to the superior border of the patella. Again, these landmarks were considered too invasive of privacy for a longitudinal study. These pairs of levels differ by less than 1.0 cm, and paired measurements at the two levels are almost identical. Measurements of gonial skinfold thickness, acromiale height and knee height are made in the Fels Longitudinal Study, although these are not among the measurements for which recommendations were made at the Airlie Consensus Conference.

The standardization and documentation of procedures are important for collaborative studies and in training new personnel. Fortunately, the latter need is uncommon. The four staff members who have close contact with the participants have worked in the study for a total of 50 years. Despite their experience, they are monitored by Dr Cameron Chumlea who joins them in the measurement of one participant each week. Doctor Chumlea is also responsible for training new anthropometrists.

Since 1929, all anthropometric variables have been obtained by a pair of anthropometrists working independently, except during the measurement of recumbent length for which they worked as a team and repeated the measurement after exchanging roles. The only landmarks marked on the participant are the levels of the mid-arm, maximum calf, and mid-thigh circumferences, and acromiale. These locations are checked by another anthropometrist. Complete replication of measurements, as at Fels, may be unique. By comparison, Goldstein (1979) describes the quality control procedures in the cross-sectional Cuban National Growth Study as elaborate and implies they are a model to be followed, although only 0.9% of the participants in this study were re-measured (Jordan *et al.*, 1975).

Until 1970, means of paired measurements were recorded but the original measurements were not retained. Children were re-measured four times if the differences between paired measurements were large, but the decisions to re-measure were not based on operational rules. Since 1970, the measurements made by both anthropometrists have been entered into a computer within the anthropometric laboratory while each participant is present. A program identifies those variables for which the differences between the observers' measurements exceed the set tolerances (maximum acceptable values; Table 2.1) or for which the mean of the paired measurements is outside the 5th to the 95th percentile range for age and sex.

When the tolerance for an interobserver difference is exceeded, or the mean is outside the 'normal range,' a computer program prompts a 'warning sign,' and both observers measure the variable again without knowing whether the reason for the repetition is a large interobserver difference or an unusual mean. When a participant is re-measured, the mean of the third and fourth measurements is used in the data analyses unless these measurements differ by more than the set tolerance or the mean of the third and fourth measurements is outside the normal range. In one or both of these circumstances, the mean of all four measurements is used. The frequency with which this occurs for each variable and for each

Table 2.1. *Set tolerances for interobserver differences in anthropometric variables*

Weight: 100 g except before 3 years (40 g)
Stature, recumbent length, sitting height, acromiale height: 1.0 cm
Arm length: 0.5 cm
Circumferences: head, thigh, arm, calf, 0.2 cm; hip, abdomen, chest, 1.0 cm
Breadths: knee, 0.2 cm; elbow, 0.3 cm; biacromial and bicristal, 1.0 cm
Skinfolds: triceps, biceps, subscapular, anterior chest, gonial, 0.2 cm; suprailiac,
     midaxillary, lateral calf, 0.3 cm

Table 2.2. *Selected distribution statistics for interobserver differences in skinfold thickness measurements for children aged 12–18 years*

| Site | Number of pairs | Technical error (mm)[a] | Coefficient of reliability (%)[b] |
|---|---|---|---|
| Triceps | 738 | 0.76 | 97.94 |
| Biceps | 739 | 0.72 | 96.02 |
| Subscapular | 738 | 0.69 | 98.38 |
| Midaxillary (vertical) | 738 | 0.93 | 96.46 |
| Midaxillary (horizontal) | 266 | 0.86 | 97.15 |
| Anterior chest | 715 | 0.99 | 95.30 |
| Suprailiac | 738 | 1.29 | 96.81 |
| Lateral calf | 722 | 0.69 | 97.49 |
| Gonial | 185 | 0.58 | 92.42 |

[a] The technical error is a measure of the extent of agreement between two corresponding values recorded for the same individuals.
[b] The coefficient of reliability is a measure of the agreement between two corresponding values recorded for the same individuals, adjusted for the difference between individuals.

anthropometrist is reviewed at fixed intervals to determine whether further re-training is indicated. This set of procedures constrains the technical errors and increases the reliability.

The first computer program used in the Fels study to check the accuracy of the recorded data also calculated the increments from the previous examinations and identified those outside the range from the 5th to the 95th percentile for the group. Children whose measurements were deemed acceptable on the basis of interobserver errors and the levels of the means were re-measured if their increments were outside the normal range. This check of increments was discontinued for several reasons. Good incremental reference data were lacking for many of the variables being measured, and the program was slow because the observed increments had to be adjusted for variations in the lengths of the intervals between examinations before they were compared with reference data. The most compelling reason for omitting this check was, however, the fact that the increments must be correct if the status values at the beginning and the end of the interval are correct, as assured by the other checks.

All the recorded data are retained in computer files except the first set of blood pressure measurements. This set is made to relax the participant and to compress the vessels prior to the measurements that will be recorded. The means of the second and third sets of blood pressure measurements are used in the analyses unless pairs of corresponding values differ by 10 mmHg or more. When this occurs, two more sets of measurements are obtained and the means of the fourth and fifth sets are used in the analyses unless, as occurs rarely, these also differ by 10 mmHg or more. On such occasions, the means of the second through the fifth measurements are used in the analyses.

Summary statistics of observer differences for all variables, are calculated within seven age groups, at 6-month intervals. These statistics relate to data recorded during the previous 6 months and also to all the data recorded since January 1976. Some of these results are presented in Table 2.2. It has been claimed that skinfold thicknesses cannot be measured reliably, but these coefficients of reliability show that, in children aged 12 to 18 years, the percentages of the total variance due to measurement errors range from only 1.62% to 7.58%. Methods for the measurement of body composition and risk factors for cardiovascular diseases are described in Chapter 7, together with the measures taken to assure that the recorded data are of high quality.

The important topic of radiation protection has received considerable attention at Fels. The radiographs have been used in studies of maturation of bones and teeth, and of body composition. In the early years of the Fels study, many radiographs were taken to assist studies of the growth of long bones and the craniofacial skeleton and to analyze the maturation of the skeleton. Therefore, great care was taken to minimize irradiation. As a result, the recorded skin doses in the gonadal region and elsewhere were too low to be meaningful (Garn, Silverman & Sontag, 1957b[168]; Garn, Silverman & Davis, 1963a[165], 1964d[166]; Garn et al., 1967a[90]). These low doses were achieved by the use of fast film, external filtration, gonadal shields and a collimator cone. Nevertheless, the radiation doses tended to increase with age and were larger for radiographs of the thorax than for the extremities. A set of radiographs, in 1964, resulted in a gonadal dose of 0.15–0.30 mrem, which should be considered relative to the irreducible background radiation from the environment of about 200 mrem/year. Currently, very few radiographs are taken.

### Data management

Data management in the Fels Longitudinal Study is designed to eliminate ambiguities in the recorded data, to facilitate the entry of data to a computer and to check relentlessly for errors at various stages of the process. Despite this, some errors escape detection. The Fels data management system can be considered a computerized filing program that also ensures that the data are of high quality.

Each questionnaire is designed to be as unambiguous as possible. The completed questionnaires are reviewed while the participants are present so that further information can be obtained if any answers are unclear. The recording forms for measurements are designed to accept data in the sequence set in the protocol. Firm adherence to a set sequence reduces the number of errors in the recorded data because measured values are less likely to be assigned to the wrong variables, and anthropometrists become accustomed to the order of the measurements.

Figure 2.3 shows the major steps in data management. At the steps in double rectangles, checks are made for possible errors by comparing printed data (print-outs) with the original records. After any errors have been corrected, the accuracy of the changes is checked on a subsequent print-out before the next step. The first step is the entry of the measurement or questionnaire data through a keyboard to a microcomputer. The entered data are listed in print-outs and checked against the hand-written records. Any discrepancies are corrected and the accuracy of these corrections is checked by comparing corrected print-outs with the hand-written records.

The data are then outputted as a sequential file which is checked and corrected as necessary before being converted to PC/SAS data format. In the next step, descriptive statistics are obtained that may lead to the identification of outlying values which are checked against the original records. Again, any necessary corrections are made and these corrections are checked on a print-out before the data are added to the master data base. Copies of the master file are maintained in several locations.

Working files that are retrieved from the master data base are fully documented. In the construction of a working file, care is taken to include all the variables that will be needed, but exclude other variables. These working files are retained on tape for at least 1 year after the results of the analyses have been published.

While it is hoped that there would not be any loss of participants from a longitudinal study, this hope is not realistic. Because loss of participants (attrition) does occur, the Fels' data are examined at 4-year intervals to determine whether the attrition has been random. If withdrawals from the

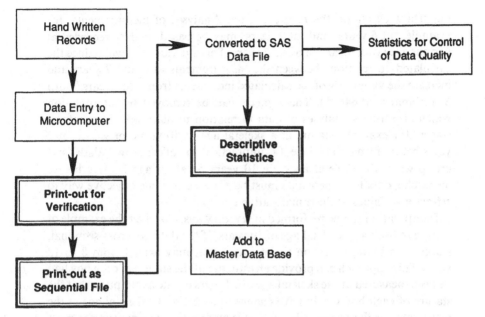

Fig. 2.3 A diagram of the steps in the management of data from their collection to their addition to the master data file.

study were more common among particular types of participants than others, the resulting bias would reduce the ability to generalize from the study findings to other groups. Participants who withdraw from the study may differ in some way from the remainder, but many of these differences, e.g., place of residence, may not affect the variables being analyzed. For the variables that have been examined, the last measurements of those who withdrew from the Fels Longitudinal Study did not differ significantly from the measurements of those who remained in the study. Therefore, the loss of participants appears to be random in regard to the measured variables and it can be concluded that the distributions of the data from the present participants are similar to those that would have been obtained had data been collected from all those enrolled.

### Interpolation

For some analyses, it is necessary to adjust the recorded data for the differences between the ages at which the participants were measured and the scheduled ages for measurement. Many small differences between these ages occur because examinations are not scheduled on Sundays or public holidays; larger differences occur when participants are sick or away on vacations (see Fig. 1.4). These variations in ages at examinations may have important effects on the recorded data. Analyses of measurements, for example, 'at 6 years' and 'at 7 years' may be based on data recorded at various ages with means of 6 and 7 years. This age variation will reduce the calculated correlations between the measurements at 6 and 7 years and increase the variability of the calculated increments from 6 to 7 years (Garn & Rohmann, 1964b[140]). These effects can be removed by interpolation which also assists analyses of data in relation to maturational events or stages: for example, stature at a skeletal age of 10 years, or weight at 2 years before menarche. Since it is impractical to collect growth data for a group when the skeletal ages are 10 years or at an age 2 years before menarche, data near these ages must be adjusted in some objective way to estimate variables at these maturational ages.

Interpolation can be performed in several ways. Consider the example of stature in boys at a skeletal age of 10 years. If the data are cross-sectional, stature can be regressed on skeletal ages that may extend from 8 to 12 years. This approach can provide an estimate of the stature of each boy had he been measured at the skeletal age of 10 years on the assumption that the stature of each boy would have changed in relation to skeletal age at the mean rate for the group. Although this assumption is incorrect for most boys, it is used commonly because it is the only way to interpolate from cross-sectional data.

Other interpolation methods are preferable when the data are serial. One simple method is linear interpolation between the values recorded at the last examination before and the first examination after the age for which the estimate is required. For example, if the stature of a boy were measured at skeletal ages of 9.4 and 10.2 years, his stature at a skeletal age of 10 years can be estimated by linear interpolation. This usually provides acceptable estimates if the pair of measurements are 1–2 years apart, but the estimates may be erroneous during periods when growth rates change rapidly. Nevertheless, this may be the best available method when a participant has only two data points.

When there are numerous serial data points for each participant, a mathematical function should be fitted to the data for the individual; the parameters of this function are used to estimate stature at a skeletal age of 10 years. In this context, a mathematical function is a set of transformed ages that are differentially weighted by coefficients and usually combined with a constant (intercept) to describe the change in the variable after the age of the intercept and to estimate the value of the variable at an age. This procedure works well if there is a close match between the fitted values and the observed data, and if there are sufficient serial data points near the age for which the estimate is required.

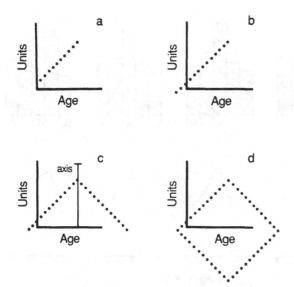

Fig. 2.4 Stages a–d in the transformation of serial data prior to fitting a Fourier function (see text for details).

Beginning in 1974, a Fourier function was used in the Fels Longitudinal Study to estimate values at target chronological ages and at selected maturational ages or events. This procedure involved transforming the data into two orthogonal components (Fig. 2.4). A set of hypothetical data is shown in Fig. 2.4a. In Fig. 2.4b, the first three points have been rotated to the left around a vertical axis passing through the first point and then rotated downwards around a horizontal axis through this point. This new data set was then rotated to the right around a vertical axis passing through the last point to obtain the data in Fig. 2.4c. Finally, the data were rotated downwards, around a horizontal axis passing through the first and last points, to obtain the enclosed area shown in Fig. 2.4d. A Fourier function was then applied to the margin of this area using a program that provided measures of the goodness of fit. This procedure was applied to all the variables that were analyzed commonly and separate files of observed and interpolated data were maintained. The routine application of this method was stopped in 1978 because it was expensive, time consuming, and the program had to be run whenever additional data points were recorded.

The Fourier method was replaced by mathematical functions which, if they fit well, can provide interpolated values based on all the serial data for an individual. Mathematical functions are now used at Fels to interpolate when there is a specific need. A mathematical function fitted to stature, for

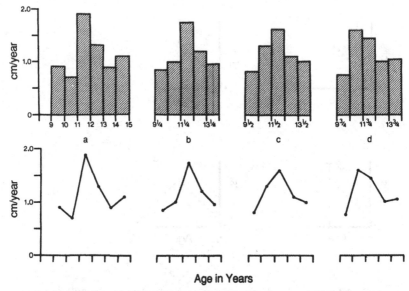

Fig. 2.5 Annual increments in bicristal diameter measured on radiographs of a female Fels participant (No. 170). The pairs of bar graphs and linear plots illustrate how variation in the ages at which annual intervals begin can alter the observations and selection of the age at the peak rate of growth. (Roche, A.F. 1974a. Differential timing of maximum length increments between bones within individuals. *Human Biology*, **46**, 145–7[178]. Redrawn with permission from *Human Biology*.)

example, will smooth the data and will assist determination of the age of occurrence of peak rate of growth (peak height velocity, PHV).

A simpler approach to the identification of the age at peak height velocity is to calculate annual increments in stature and record age at PHV as the midpoint of the annual interval with the largest increment. This approach can lead to large errors, as demonstrated by data for annual increments in bicristal (hip) diameter in one Fels girl (Fig. 2.5). The pairs of matching bar and linear graphs indicated by 'a' show annual increments (cm/year) for intervals beginning at birthdays. There would be marked differences in the increments if the annual examinations had been made 3, 6, or 9 months after each birthday (pairs b, c, and d, respectively) assuming that the growth rate was constant during each annual interval. Whereas the midpoint of the interval with the maximum increment is 11.5 years for a, it varies from 11.25 to 12.0 years among b, c, and d.

### Derivation of variables

Within the Fels study, some statistical analyses are based on variables derived from the observed data. There are two groups of derived variables from the viewpoint of data management. Monthly calculations are made of those that are analyzed commonly and are easy to compute; the calculated values are added to the permanent computer file. Those that are analyzed less frequently and are more difficult to compute are calculated only when needed.

The derived variables that are regularly added to the permanent file include the following.

(i)     Weight/stature$^2$ (W/S$^2$). This is related to total body fat and to the percentage of the body weight that is fat (percent body fat; Roche *et al.*, 1981b[265]).

(ii)    Muscle and adipose tissue areas of the arm and calf. These areas are calculated from a circumference and a skinfold thickness at the same level, assuming the cross-sectional area of the arm or calf is circular. For example, arm muscle area (cm$^2$) = [arm circumference $-$ 3.14 (triceps skinfold thickness)/10]$^2$/12.56. Additionally, the total area of the arm or calf is easily calculated from its circumference and the adipose tissue areas can be obtained as the differences between the total areas and the muscle areas. Although these calculated values are only indices of the true areas, they assist predictions of the total amounts of fat and fat-free mass in the body (Himes, Roche & Webb, 1980; Guo, Roche & Houtkooper, 1989a).

(iii)   Body density (BD). Body density is calculated from weight-in-air, weight-in-water, water temperature and residual volume. The last-mentioned is the amount of air remaining in the lungs at maximal expiration. The calculation of body composition variables, such as total body fat and fat-free mass, from body density is generally regarded as the best available procedure that is applicable to the living (Roche, 1984a[196], 1985a[199], 1987a[204]).

(iv)    Stature$^2$/resistance (S$^2$/R). The resistance of the body to the passage of a small alternating electric current can help predict the lean mass of the body (fat-free mass) and the total amount of water in the body (total body water; Kushner & Schoeller, 1986; Guo *et al.*, 1987a; Guo, Roche & Chumlea, 1989b). Resistance is useful for this purpose only when it is used in combination with an index of the length and volume of the conductor; stature$^2$ is effective as such an index. The accuracy of this approach is increased further when resistance and stature[2] are combined with reactance ($X_c$). Reactance, which is measured with the same equipment as resistance, is the inverse of capacitance. Capacitance is the storage of voltage by a condenser for a brief moment in time.

(v)    Relative skeletal age. This is calculated as (skeletal age − chronological age) or as skeletal age/chronological age. These values are preferred to observed skeletal ages because they are less dependent on chronological ages.

Other derived variables added to recent working files include the following.

(i)    Phase angle ($\Phi$). As part of bioelectric impedance studies, the phase angle ($\Phi$) may be calculated from resistance (R) and reactance ($X_c$) as:

$$\Phi = atan\ (X_c/R)$$

and converted to degrees by multiplying by 57.297.

(ii)    $W^{1.2}/S^{3.3}$. This is the weight–stature ratio that is maximally related to percent body fat in the Fels data (Abdel-Malek, Mukherjee & Roche, 1985; Roche, Abdel-Malek & Mukherjee, 1985[210]).

(iii)    Ages at peak height velocity estimated from mathematical functions fitted to the serial data for each participant.

(iv)    Parent-specific values for recumbent lengths and stature. Since tall parents tend to have tall children, and the reverse, it is desirable to adjust observed recumbent lengths and statures for the statures of the parents before performing some analyses. These procedures should be applied also in the clinical evaluation of children with unusual recumbent lengths or statures and in some comparisons between groups (Himes, Roche & Thissen, 1981; Himes *et al.*, 1985).

(v)    Weight-for-stature. Because weight and stature are correlated, weight should be adjusted for stature in some evaluations of children and in some statistical analyses. These adjustments can be made in several ways. One simple procedure, that is applicable on an age-independent basis before pubescence, is to compute percentile levels of weight-for-stature corresponding to the values observed in a large survey (Hamill *et al.*, 1977, 1979).

### Transformation of variables

Some recorded variables should be transformed before they are analyzed statistically. These transformations may be mandated by the nature of the data and by the statistical test to be applied. Tests of the significance of differences between groups for one variable (e.g. t-test) or for several groups simultaneously (analysis of variance) require that the data be normally distributed. This requirement was recognized early in the Fels study (Sontag & Wallace, 1935a[305]). Problems arising from the non-normality of the distributions may be circumvented by using a distribution-free test. The specific question being examined will determine whether a distribution-free test, or normalization of the distributions, is indicated.

For some variables, such as skinfold thicknesses and ages at onset of ossification, skewed distributions are common but not universal (Garn *et al.*, 1965g[71]). Such a distribution can be normalized using a Box–Cox transformation or logarithmic (log) transformation (Edwards *et al.*, 1955; Box & Cox, 1964; Patton, 1979). Log transformations may be appropriate also in the presence of multiplicative relationships in the data, as may occur between the independent variables in a predictive equation. In these circumstances, log transformations simplify the demonstration and analysis of the multiplicative relationships. Finally, log transforms of one or both variables commonly change a curvilinear bivariate relationship to a rectilinear one that is easier to interpret. This occurs for some ratios used to describe fat patterning. Some have suggested the use of Box–Cox (1964) transformations to develop reference percentile curves. This approach has been criticized by Roche and Guo (1988)[237].

## Statistical analyses in the Fels Longitudinal Study

This section presents an overview of the methods that have been applied to analyze data in the Fels study. More detailed descriptions of any unusual methods will be given in the parts of this volume that present the substantive results of the analyses. Additionally, this section will direct attention to improvements in statistical methods for which Fels investigators, or colleagues working with Fels data, have been responsible. Generally, the statistical methods applied were appropriate for the questions to be answered or the hypotheses to be tested, but some exceptions occurred early in the study when statistical tests were applied to some samples that were too small to provide conclusive results.

Some statistical methods that were developed at Fels for specific purposes have had considerable influences. Examples are the excellent review of statistical methods by Garn (1958b[53]) that improved the statistical analyses of a whole generation of orthodontists and papers that have had a major impact on the statistical analyses made by pediatric nephrologists (Roche, 1978a[184], 1978b[185]; Potter *et al.*, 1978).

One common statistical procedure is the calculation of descriptive statistics such as the mean and standard deviation (SD). Commonly, growth data, particularly those related to body fat and those for increments, are not normally distributed (Garn, Rohmann & Robinow, 1961c[152]; Garn & Rohmann, 1964b[140]). Therefore, skewness and kurtosis have been evaluated and, when indicated, percentiles have been used to describe the distributions, and the significance of differences between non-normal distributions for group means has been tested by the non-parametric Mann–Whitney test, or the distributions have been transformed.

When distributions were skewed, some consideration was given to possible causes for this skewness. Severe error bias is unlikely, but truncation and mixtures of normal distributions are possible. Truncation occurs, as pointed out by Garn and his co-workers (1961b[144]), in the distribution of stature increments after 17 years because some will have reached adult stature at 17 years. Other distributions may be non-normal because several normal distributions are combined in the data set. A family of distributions with location and scale factors was developed that can be fitted to a set of data (Mukherjee, 1982; Mukherjee & Siervogel, 1983; Mukherjee *et al.*, 1984). In addition, a set of multimodal distributions has been developed for use when a mixture of normal distributions is untenable (Mukherjee & Siervogel, 1983). These analyses of distributions can be valuable in genetic studies. In addition, Mukherjee and Hurst (1984) described procedures that facilitate descriptions of the distributions of discrete or continuous data. In other statistical work, a curve-fitting procedure was developed that provides a normal approximation to a binomial distribution (Lee & Guo, 1986). This facilitates the testing of hypotheses using data from large samples.

Hypotheses relating to differences between groups have been tested in many analyses made at Fels. Severe methodological problems are not present if the test is for one variable and it relates to the difference between the means for boys and girls. Caution is necessary, however, when testing the significance of differences between, for example, the means for head circumference in 4-year-old boys and 5-year-old boys. When the data come from a longitudinal study, they will be biased due to the inclusion of many of the same boys in both age groups. If only a few were measured at both ages, which is unusual in data from the Fels study, it might be preferable to exclude these and thereby make the data independent at the two ages. Sometimes, multiple measurements of individuals can be an advantage. For example, such data allowed a multivariate analysis of variance for repeated measures to estimate the examination effects in serial measures of hearing ability (Roche *et al.*, 1983a[255]).

During the analysis of Fels data, regressions have been calculated to describe the relationship between a variable and age or between two or more variables at an age. The simplest type of regression is a linear model. As an example, a linear model can be used to describe the relationship of weight to stature in 6-year-old boys, as shown in Fig. 2.1. The regression equation (weight in kilograms $= 30.915 + [0.446 \times$ stature in centimeters]) summarizes this relationship and allows predictions of weight from stature. The 95% confidence limits of this regression, which are included in Fig. 2.1, allow the identification of outlying data points. In some analyses, one variable has been regressed on a logarithmic transformation of another variable.

Multiple regression has been used commonly at Fels in the development of predictive equations that estimate one variable from a combination of other variables. This is used in the Roche–Wainer–Thissen method for the prediction of adult stature from variables observed during childhood (Roche, Wainer & Thissen, 1974a[270], 1974b[271], 1975a[272], 1975b[273]).

Typically, predictive equations are derived by least squares regression in which the sum of the squares of the residuals is minimized. In this sense, 'residuals' are the differences between the fitted values and the observed points. This approach provides the best equation for the given data but, if the multiple predictor (independent) variables are significantly correlated (multicollinearity), the coefficients in the predictive equation are likely to be unstable. In the presence of this instability, predictive equations may perform poorly when applied to other samples. This is a major problem because such equations are, of course, developed so that they can be applied to samples other than those from which they were derived.

The severity of the problem can be determined by calculating the variance inflation factor for each independent variable included in a predictive equation. The variance inflation factor $(1/1 - R^2)$ reflects the extent to which each independent variable can be predicted from the other independent variables in combination; values greater than 10.0 are generally accepted as indicating potentially serious multicollinearity.

The solution to the problem is not to abandon the least squares approach but to modify the data from which the predictive equation is derived. Ridge regression should be considered for this purpose. In ridge regression, the interrelationships between the independent variables are reduced by adding a small constant to the diagonal elements of the variance–covariance matrix of the independent variables.

The constant employed is the smallest value beyond which further increases in the constant would have little effect on the coefficients in the predictive equation. Unfortunately, these possible values for the constant differ slightly from one independent variable to another, making it difficult to select the optimal value for the constant in this way. Alternatively, the value can be determined precisely using a PRESS procedure in which one data point is omitted at a time, after which it is estimated from the remaining information. The differences between these estimates and the omitted data are the PRESS residuals; the sum of the squares of these residuals is called the PRESS statistic, and the value chosen for the constant is that at which the PRESS statistic is smallest.

The PRESS procedure also identifies data points with large residuals. These points, that were estimated poorly by the data remaining after the omission, one at a time, of values for independent variables, are called leveraged observations because they make large contributions to the instability of predictive equations. Their effects are reduced in robust estimations because leveraged observations are given smaller weights than other data points when predictive equations are being developed. The equations developed by robust estimation perform better than equations derived from the original data when both are applied to other samples (Roche, Guo & Houtkooper, 1989b[241]).

Piecewise regressions have been used at Fels to identify the ages at which changes occur in the patterns of growth (Roche & Davila, 1972[222]; Roche, Davila & Mellits, 1975c[227]). Using this procedure, the ages of cessation of growth in stature and the ages at which increases in weight change from being curvilinear to rectilinear have been estimated for individuals. For this purpose, a pair of mathematical functions was fitted to the serial data for each participant from the age at peak height velocity to 28 years, and the junction between the two curves was changed by one data point at a time. The goodness of fit for the total data set was determined for each junction, and the junction (age) at which the goodness of fit was maximal was selected as the age at which the pattern changed for the individual. Although the procedure worked well, particularly for stature, a single function would be fitted in a future analysis of this topic and the function would be integrated to estimate ages at critical points in the growth pattern.

Logistic regression has been used to predict a dichotomous outcome, for example the absence or presence of a value greater than a 'cut-off' level. In logistic regression, the probability (P) of the outcome is divided by $(1-P)$. This mathematical expression $(P/1-P)$ is called the odds ratio; the logarithm of this (log odds ratio) is used as the outcome variable. Logistic regression has been applied to Fels data when the outcome variable was the presence or absence of 'overweight' in adulthood, and the independent (predictor) variable was the corresponding measurement at various ages during childhood (Roche & Guo, 1987[236]; Roche, 1987c[206]; Siervogel, Guo

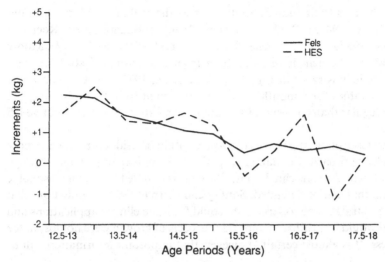

Fig. 2.6 Median values for 6-month increments in weight for girls from the Fels study and from the US Health Examination Survey (HES) conducted by the National Center for Health Statistics, at ages from 12.5 to 18.0 years. (Roche, A.F., Davila, G.H. & Mellits, E.D. (1975c). Late adolescent changes in weight. In *Biosocial Interrelations in Population Adaptation*, ed. E.S. Watts, F.E. Johnston & C.W. Lasker, pp. 309–18[227]. Redrawn with permission from Mouton Publishers.)

& Roche, 1991). Corresponding analyses have been made of serum lipid levels (Guo, Chumlea, Siervogel & Roche, 1991). Logistic regression has been used also to test hypotheses that specific genetic markers are linked to particular phenotypes (Falk *et al.*, 1982).

The central purpose of any longitudinal study is to describe change in individuals. The simplest mathematical description, both conceptually and computationally, is the calculation of increments between successive examinations. Each increment must be adjusted so that it matches the change during a fixed interval that is typically 6 months or 1 year. These adjustments are necessary because it is highly unlikely that participants in a longitudinal study will be examined at intervals of exactly 6 months or 1 year.

Increments describe growth rates during particular intervals but they provide little information about growth patterns and they can be affected markedly by measurement errors. Additionally, the ages at measurement will not coincide with critical points on the growth curves (Roche, 1980a[190]). Because increments require only two data points per individual, they can be readily applied by clinicians and others. Reference data for increments during 1-month and 6-month intervals have been published from the Fels Longitudinal Study to assist the evaluation of growth rates (Robinow, 1942a; Roche & Himes, 1980[247]; Baumgartner, Roche & Himes, 1986a; Guo, Roche & Moore, 1988; Roche, Guo & Moore 1989a[242]). The benefit of using data from a longitudinal study for this purpose is illustrated in Fig. 2.6 (Roche *et al.*, 1975c[227]). This shows the median values for 6-month increments in weight from the Fels Study are more regular than corresponding data from a nationally representative US study.

The description of patterns of growth within individuals requires a more complex approach. Fundamentally, the problem is to draw a line through a set of data. Anyone can do this. The issue is to base the line on reasonable criteria that can be defended. Sontag and Garn (1954[285]) pointed out that fitting a straight line to serial data could facilitate clinical applications and that it would be justified if important aspects of growth were not lost in the process. It is almost certain, however, that important information will be lost.

Typically, the changes in a variable are analyzed in relation to age. These changes must be described by a mathematical function that summarizes the pattern of growth using a few numbers that are called the coefficients of the parameters in the fitted mathematical function. It is not easy to choose the best function because the patterns of change in individuals tend to be more complex than those in group data and they vary among individuals.

The most direct benefit of fitting a mathematical function to serial data for individuals is the description of change. To be useful, this description must be considerably more parsimonious than the observed data. Expressing this another way, the number of mathematical terms in the function must be considerably less than the number of data points. Ideally, each term in the model will be biologically interpretable, e.g., rate of growth, change in rate of growth, but commonly, such interpretations are uncertain due to correlations between the coefficients of the model. In this area, as in so many others, interrelationships are the rule, not the exception. As stated poetically by Francis Thompson: 'Thou cans't not stir a flower without troubling a star.'

Since the chosen model will have fewer parameters than the number of data points, some smoothing will result. This is desirable because many changes in observed serial data result from measurement errors or small fluctuations that are not biologically informative, e.g., fluctuations in weight due to variations in the sizes of previous meals. If there were many parameters in the function, there would be little smoothing and the estimates of the coefficients of the parameters may be unstable (large confidence limits), thereby complicating biological interpretations and statistical inferences.

The function can have too few parameters. A very simple function will smooth excessively with the loss of important information. The choice of an equation that provides a parsimonious solution, retains critical information, fits well, smooths moderately, and can be readily integrated requires considerable judgment and experience on the part of a biostatistician. This approach is not recommended for clinical purposes. Mathematical functions are fitted to serial data to serve research purposes; many data points are required and the approach is retrospective in nature.

Early steps in the description of growth patterns are the selection of the features that are critical for testing particular hypotheses, e.g., age at peak height velocity, and the determination of the general nature of the serial changes by visual examination of plots for some individuals chosen at random. An experienced biostatistician can then derive a reasonable set of functions and test them for goodness of fit. The patterns of the residuals in relation to age may show that the functions need to be modified. Large correlations between successive residuals could show that the chosen model is not providing an accurate description of the growth patterns (Bock & Thissen, 1980).

Functions that describe growth patterns can assist comparisons between individuals and between variables, the prediction of adult status and evaluations of the effectiveness of therapy (Roche, 1971a[174]; Holm *et al.*, 1979). In the best of circumstances, the description provided by a function is an approximation because one cannot be sure that all the errors and only the errors have been removed by the smoothing process. One takes heart from the attitude of Tukey: 'An approximate answer to the right problem is worth a good deal more than an exact answer to an approximate problem.'

Some flexible functions have been fitted to serial data from the Fels Longitudinal Study but usually fixed functions have been used. Flexible functions do not have predetermined shapes and, therefore, they are suitable for variables, such as skinfold thicknesses, for which the growth patterns differ among individuals. Longitudinal principal components, splines and kernel estimation are examples of flexible functions. Fixed functions are appropriate when the growth patterns have the same shape for all individuals even though they differ in the timing and the amounts of change. Logistic functions are examples of fixed functions that fit well to variables such as stature and head circumference from birth to 6 years.

Longitudinal principal components analysis (LPCA) is one type of flexible model that has been used to describe growth patterns. This method is a variant of principal component analysis which is usually applied to an intercorrelation matrix for multiple variables at one age while LPCA is applied to an intercorrelation matrix for one variable at multiple ages. This procedure yields a set of components that vary with age but are common to all individuals. Each participant has a specific positive or negative age-invariant coefficient for each component. To some extent, the components can be interpreted biologically and their coefficients can be analyzed statistically. This method has been used in the Fels Longitudinal Study to describe changes in recumbent length, weight/stature$^2$ and skinfold thicknesses, although it requires a complete data set for each individual (Roche, 1971a[174]; Cronk *et al.*, 1982a, 1982b, 1983a).

A new flexible function for the analysis of serial data, that is robust to missing values, has been developed (Guo *et al.*, 1987b) and is being applied in cancer research and drug trials. In this method, serial data for individuals are pooled between groups receiving different drugs or different dosages, and the data are ranked at each age. These ranks are summed for each group and the significance of the differences between the groups can be tested.

A cubic spline has been used to smooth empirical percentile values for weight, recumbent length and head circumference that were obtained by cross-sectional analyses of Fels data (Hamill *et al.*, 1977, 1979). On other occasions, low-term Fourier analysis has been used to smooth empirical percentiles and serial data for individuals (Roche & Himes, 1980[247]; Roche, 1980a[190]; Cronk & Roche, 1982).

More recently, kernel estimation has been used to smooth data for blood pressure and other variables (Guo *et al.*, 1988, 1989c). Kernel estimation is based on weighted averaging of the observed values within specified age intervals (Guo, 1990). The lengths chosen for these intervals determine the extent of smoothing and the goodness of fit; short intervals lead to less smoothing but the fits are better. The weights given to the values within each age interval vary in relation to the differences between the ages at examinations and the midpoints of the age intervals. The data points that are more divergent from the midpoints are given lower weights. A procedure for calculating the confidence limits of kernel estimates has been developed (Guo, Siervogel & Roche, 1990).

In a comparative study, kernel regression, the triple logistic model, and the Preece–Baines model were fitted to serial data for stature (Guo *et al.*, in press). The Preece–Baines model did not describe the mid-growth spurt, which occurs at about 6 years, and the parameters from this model showed an earlier onset and longer duration of the pubescent spurt with a less rapid increase in velocity than either kernel regression or the triple logistic model. The kernel regression and triple logistic methods provided similar descriptions of the pubescent spurt, but the estimates from kernel regression showed an earlier onset and a more rapid increase in velocity for the mid-growth spurt than did the triple logistic model.

A need arose to smooth, across age, the coefficients in numerous age-specific multiple regression equations developed for the prediction of adult stature from childhood variables (Roche *et al.*, 1974a[270], 1974b[271], 1975a[272], 1975b[273]). After this smoothing, a function was fitted to the smoothed data and interpolations were made to intermediate ages. The new statistical method developed for this work took the interrelationships between the variables into account (Roche *et al.*, 1975a[272]; Wainer & Thissen, 1975).

This multivariate smoothing was accomplished after transforming the coefficients so that they became independent of each other (Björck, 1967). Initially, they were smoothed by the '53h' method of Tukey (1972) in which running medians are obtained of each successive five points and then of each successive three points. This provided two smoothed estimates for each age; the means of these pairs were subtracted from the observed values and the same procedure was applied to the residuals. These steps were repeated until further changes did not occur in the smoothed values. After this initial smoothing, a polynomial function was fitted to the smoothed values and these polynomial functions were transformed back to the original variables.

Time series has not been used often in the analysis of Fels data but it was applied to determine whether patterns of change in bioelectric impedance (resistance) were related to timing within the menstrual cycle, assuming a first-order autocorrelation (Roche, Chumlea & Guo, 1986a[217]).

A three-parameter polynomial function, that was used to describe serial changes in weight and recumbent length during infancy (Kouchi, Mukherjee & Roche, 1985a; Kouchi, Roche & Mukherjee, 1985b), included a constant term (intercept), a scale term (slope) and a power term (change in rate of growth). The function fitted well to the data, which is essential but, in many analyses, it is also important that the parameters of the model be biologically interpretable. Such interpretations are difficult, if the parameters are significantly intercorrelated. In these analyses by Kouchi and her colleagues, the total intercorrelation matrices of the parameters were adjusted for the within-individual correlations, using a procedure introduced by Bock *et al.* (1973) in an earlier analysis of Fels data. The new matrices obtained for the within-individual correlations provided more stable parameter estimates for individuals and facilitated comparisons between individuals.

A four-parameter polynomial function has been used to describe growth in head circumference from birth to 18 years (Roche *et al.*, 1987b[258]; Guo, Roche & Moore, 1988) and three-parameter polynomials have been applied to logarithms of weight/stature² and to weight and recumbent length during infancy (Siervogel, Mukherjee & Roche, 1984; Roche, Guo & Moore, 1989a[242]). Since the function fitted well to the data for infancy, and because observed data at 1-month intervals are scarce, the function was used to estimate values for status at 1-month intervals and for increments during 1-month intervals for weight and recumbent length from birth to 1 year.

Fixed functions have been developed to describe growth in skeletal dimensions (Bock *et al.*, 1973; Thissen *et al.*, 1976; Roche *et al.*, 1977a[251]). The general patterns of growth in these variables are similar among individuals despite differences in intercepts, rates of change, and the timing of critical events. A double logistic function was developed to describe growth in recumbent length from 1 to 18 years within individuals (Bock *et al.*, 1973). In this function, the first logistic component describes pre-pubertal growth and continues at a low level to 18 years. The second logistic component describes the adolescent growth spurt. Although the fit is generally good, the function tends to over-predict from 4 to 6 years and to under-predict slightly in early pubescence and late adolescence. Only three, or at most four, of the six parameters in the function are needed to describe individual differences. Publication of this function sparked a sudden increased interest in fitting functions to growth data.

Later, Bock & Thissen (1976, 1980) developed a triple logistic function to provide a better description of growth from 1 to 18 years, and used it to characterize the unusual patterns of growth in some Fels participants (Bock, 1986). This function fitted better than the double logistic function and it defined the mid-growth spurt clearly. Bock & Thissen also described elegant methods for adjusting the correlations between parameters. Different fixed models were used to describe individual patterns of growth in weight and recumbent length from 3 months to 6 years and changes in weight/stature² from 2 to 18 years (Siervogel *et al.*, in press; Byard, Guo and Roche, 1991).

Lestrel (Lestrel & Brown, 1976; Lestrel & Roche, 1984, 1986) extended mathematical modelling to the description of two-dimensional shapes using Fourier series analysis. Radii were drawn from a centroid to landmarks on the outline of the shape, after which a polynomial with Fourier terms was used to describe the lengths of the radii. This description of shape was made independent of size. The method has been applied to the silhouette of the cranial vault and paired ectocranial and endocranial outlines to measure cranial thickness (Lestrel & Roche, 1977) and to outlines of nasal bones (Lestrel, Engstrom & Bodt, 1991).

Principal components analysis has been used to describe the patterns of adipose tissue distribution. The aim of these cross-sectional analyses was to identify primary dimensions of patterns in the anatomical distribution of adipose tissue that are independent of overall body fatness (Baumgartner *et al.*, 1986b). In addition, principal component analysis has been used, in the development of predictive equations, to investigate the intercorrelation matrices and assist the selection of independent variables from many candidates (Roche *et al.*, 1975a[272]). Cluster analysis of the variables has been used for the same purpose (Roche *et al.*, 1975a[272]; Baumgartner, Siervogel & Roche, 1989b).

In principal component analysis, a reduced set of linear combinations of variables is sought that provides essentially the same information as the original variables. If two or more variables within a principal component have closely similar loadings, then only one member of such a set need be retained for further analysis. Scores for individuals can be calculated from the loadings for each component creating new variables that are independent. These new variables retain all the crucial information and they can be used in multivariate analyses without fear of multicollinearity. In principal components analysis, the potential independent variables are grouped and, if there are high correlations between the group members, only one member of a group is retained for further analysis. Alternatively, the groups themselves may become the object of analysis.

Profile analysis has been used to analyze patterns of adipose tissue thicknesses and ages at onset of ossification in the Fels study using both absolute and relative values (Sontag & Reynolds, 1944[297]; Sontag & Lipford, 1943[288]; Garn, 1955a[44]). These profiles have been compared visually and they have been analyzed statistically in relation to genetic hypotheses using either correlations between corresponding values for pairs of related individuals or the standard deviations of the differences among values.

Robinow (1968) wrote an interesting paper relating to sample size determination and described simple statistical procedures that can be applied in the field. He suggested that data for weight and recumbent length could be combined for the two sexes during infancy, thus reducing the sample size required. It has been shown that this is not justified for recumbent length (Roche & Guo, unpublished data). Robinow noted that chi-square tests can be used in the field to determine whether groups differ in the percentage under the 5th percentile and he described a method by which data for weight from 1 month to 5 years can be adjusted to an 'equivalent' weight at a central age because there is little change in the shapes of the distributions during this age range. This approach could be helpful if differences in growth between shorter age intervals are not obscured.

The preceding discussion provides an overview of statistical applications in the Fels Longitudinal Study. The wide range of purposes and the consequent variety of analytic methods used, even within topic areas, will become more evident in the later chapters. The methods applied recently are more elegant and effective than those applied in the early years of the study, but we have not reached the end of these advances. Further improvements in data management and statistical methods must occur and they will lead to increased understanding of the phenomena being studied.

# 3 *Prenatal, familial and genetic studies*

'The growth of the fetus increases more and more, in equal time, till it escapes the womb.'

*George LeClerc Buffon* (1707–1788)

The original mission of the Fels Research Institute included the serial study of individuals before birth; this aspect of the mission has not been neglected. Particularly during the early years of the Fels Longitudinal Study, strenuous efforts were made to perform prenatal studies and successes were achieved although the available technology allowed only a narrow range of investigations. Better methods are now available that could assist prenatal studies of growth, maturation and body composition, but some involve radiation (computerized tomography), and others are expensive (ultrasonography, magnetic resonance imaging). Imaging procedures have not been applied serially in normal pregnancies although these studies have great potential.

Some investigations made within the Fels Longitudinal Study that relate to the fetal period are described with physical growth (Chapter 4) and skeletal and dental studies (Chapter 6). The prenatal investigations described in this Chapter have been grouped under the headings: prenatal studies, and familial and genetic studies.

## Prenatal studies
### Diet and nutrition
During the 1930s and early 1940s, the relationships between prenatal maternal diets and the size of the infant at birth were studied. This was a Herculean task. Daily dietary records were kept by 205 mothers for 4 to 7 months. These mothers were not given dietary advice; this was a 'natural experiment,' as is generally true for the observations made in the Fels

Table 3.1 *Coefficients of variation* ($CV^a$; *%*) *for creatinine and creatine in 24-hour urine samples measured daily during pregnancy (from Seegers & Potgieter, 1937)*

| Participant | Number of days | Creatinine | Creatine |
|---|---|---|---|
| III | 7[b] | 6.9 | 31.8 |
| VI | 7[b] | 0.7 | 33.5 |
| X | 8[b] | 7.0 | 51.8 |
| IX | 7[c] | 17.1 | 52.2 |
| VIII | 5[c] | 11.9 | 52.3 |
| XIII | 26 | 33.6 | 184.7 |
| XIV | 21 | 29.5 | 130.4 |
| XIV[d] | 21 | 40.3 | 120.2 |

[a] CV is the coefficient of variation which equals standard deviation/mean.
[b] 28-day intervals.
[c] Successive days.
[d] Diet rigorously controlled.

Longitudinal Study. These, and other dietary data, provided some important information.

Interest in protein intake and metabolism during pregnancy was stimulated by findings from animal experiments at the Fels Research Institute (Seegers, 1937a). These studies showed that the offspring of pregnant albino rats fed a nitrogen-free diet had low nitrogen values per unit body weight during pregnancy and low birthweights. At the time this study was published, many considered that a low protein diet during pregnancy did not affect the fetus although contrary findings had been reported from the Harvard Growth Study. It was also shown that nitrogen retention in a young primipara was similar to that reported for older pregnant women (Seegers, 1937b).

In an extension of this work, Fels data for daily intakes of protein were averaged for individuals and used to place the mothers in one of five groups (Sontag & Wines, 1947[308]). After excluding data for pregnancies resulting in premature or multiple births and those in which pathological conditions occurred in either the mother or the infant, correlations between group membership and values for weight and recumbent length at birth were not significant. This implies that homeostatic mechanisms maintain the blood levels for protein and that these levels fall only when the dietary reductions are drastic, and interfere with the supply of nutrients to the fetus.

The difference between these results and those from the Harvard study may have been due to the generally lower protein intakes in the Harvard study and their failure to exclude toxemic mothers who were treated with low protein diets. These and other findings were included in a critical review published by Sontag in 1941.

Earlier, Seegers and Potgieter (1937) reported the 24-hour excretion of creatinine and creatine in the urine of pregnant women during the third day of a uniform diet. These measurements did not change with gestational age, but they fluctuated irregularly even for the participant (Number XIV) whose diet was 'rigorously controlled' (Table 3.1). The correlations between the intake and the urinary excretion of nitrogen were also variable; the correlations within participants ranged from r = −0.5 to +0.4 for creatinine and from r = −0.7 to +0.2 for creatine.

Other studies concerned the intake of calcium and phosphorus during pregnancy. It was shown that, in Wistar rats, very low levels of vitamin D and calcium in the maternal diets led to low values for weight, and for the content of calcium and phosphorus in the offspring at birth, but the calcium:phosphorus ratio was high (Sontag, Munson & Huff, 1936[289]). These animal studies were conducted in concert with studies of calcium and phosphorus in Fels mothers. A significant correlation was reported between the serum calcium of mothers during pregnancy and that of their infants soon after birth, but many other correlations between maternal and infant variables were not significant (Sontag, Pyle & Cape, 1935[296]). Sontag and his associates concluded that maternal dietary reports were too unreliable for such investigations and that future studies should be based on weighed food portions. This decision was not implemented in the Longitudinal Study.

In partial response to this conclusion, however, Pyle and Huff (1936) described a procedure in which fixed intakes of calcium and phosphorus were maintained for 3 days. They considered this would stabilize conditions sufficiently to allow studies of calcium and phosphorus. The 3-day method was used at Fels for logistic reasons and also because it was argued that: 'One needs as short a balance period as possible if he is studying metabolism during pregnancy when the rates of change in both the fetal and maternal organism are not constant.' This is doubtful at best, particularly when the 3-day periods used in the Fels study were separated by 28-day intervals. To 'validate' the 3-day records, it was shown that the variability of calcium, protein, and nitrogen in the food, feces and urine from such records was similar to the variability reported from studies in which fixed intakes were maintained for longer periods (Pyle & Huff, 1936; Sontag & Potgieter, 1938[293]). Data from 3-day and longer periods on fixed intakes were not compared within individuals.

Applying their 3-day method to Fels mothers, Huff and Pyle (1937) identified periods of positive and negative nitrogen balance and for calcium and phosphorus jointly. They showed differences in the fecal excretion of calcium and phosphorus between those in positive and those in negative balance but not in the ingestion or urinary excretion of these minerals. If such a study were conducted now, residence in a metabolic ward would be considered mandatory and the period of fixed intake would be substantially longer.

The Fels qualitative method of recording dietary intakes and calculating nutrient intakes was compared with a quantitative method in six women who kept dietary records for a combined period of 30 weeks (Sontag, Seegers & Hulstone, 1938[301]). In the qualitative method, food diaries were kept for a week. The number of servings for each food group was used in combination with the average serving size to calculate nutrient intakes from food composition tables. In comparison with estimates from a quantitative method, the qualitative estimates were low, but the paired estimates were correlated with coefficients ranging from $r = 0.4$ for fat to $r = 0.8$ for protein. The authors concluded that the qualitative method was suitable for group comparisons but 'obviously, the method is not applicable as a part of balance study technique.' The conclusion was justified.

In 1938, Pyle, Potgieter and Comstock studied nine pregnant women using 3-day balance periods (two to six periods per women). These mothers were told how to weigh their food and instructed to eat the same kinds and quantities for three successive days. On the third day, the food was sampled and 24-hour collections were made of urine and feces. On the fourth day, blood was obtained to measure serum calcium, and basal metabolic rate (oxygen consumption) was recorded. There were significant correlations between serum calcium and calcium balance ($r = -0.4$), serum calcium and fecal calcium ($r = 0.8$) and calcium balance and gestational age ($r = 0.6$). Many correlations were calculated but they were not adjusted for the effects of multiple comparisons. Consequently the significance levels may be lower than they appear. In this study, the calcium balance data were obtained for short periods without sufficient control of intake and losses. Furthermore, the unequal loading of the women, due to between-subject variations in the number of study periods, would have led to some confounding. Better studies could be made now, but they are likely to occur in departments of nutrition where it is hoped the efforts of the pioneers are remembered.

It was noted by Sontag (1944a[279]) that some obstetricians attempt to reduce birthweight, and thereby reduce the incidence of obsterical complication, by limiting the maternal caloric intake during pregnancy. Sontag considered that this reflected a lack of concern about the possible effects of the maternal diet on the fetus. Furthermore, he postulated that the fetus may be affected by maternal emotions which can elevate the circulating levels of some chemicals that pass through the placenta. He noted that the infants of mothers with severe emotional stress during pregnancy tend to be hyperactive and irritable after birth.

### Physiology

Despite limited equipment, judged by present standards, the early Fels staff made valiant efforts to study fetal physiology (Sontag, 1944a[279], b[280]). One set of these studies related to fetal heart rates. It was noted that these rates increased significantly with maternal smoking and it was concluded that this was probably due to the passage of toxins into the fetal circulation (Sontag & Wallace, 1935a[305]). In this impressive study, the reliability of counting the heart rates was established, partly by having two observers count simultaneously. In one investigation, the fetal heart rates were determined from 5 minutes before until 14 minutes after lighting a cigarette; this counting was done 81 times in five women. There was a mean increase in rate of seven beats/minute except in one participant who was a non-smoker and did not inhale. This was one of the very earliest studies that reported an effect of maternal smoking on the fetus. Later, Sontag and Richards (1938[300]) again observed effects of maternal smoking on the fetal heart rate but, because of the marked variability and the small sample, they were unable to demonstrate statistically significant changes.

In a series of studies, it was shown that the fetal heart rate increased by an average of 12 beats/minute when a vibrating frequency of 120 beats/minute was applied to the maternal abdomen or the mother was exposed to an air-conducted tone (Sontag & Wallace, 1934[304], 1936[307]; Bernard & Sontag, 1947). These increased rates occurred independently of fetal movements and were not noted until the last trimester of pregnancy. Sontag and his colleagues suggested that these effects were associated with the innervation of the heart by sympathetic fibers or with maturation of the fetal adrenal gland.

Sontag and Richards (1938[300]) published an excellent review of the maturation of the physiological mechanisms that might regulate fetal heart rates. They realized their studies of fetal heart rates were limited because 'our organism is not seen, nor scarcely felt nor heard.' Nevertheless, they persevered and used the best methods available without concealing their defects. As they wrote: 'No amount of calibration or recalibration will reduce certain sources of error inherent in the situation.' This perseverance was dictated by strict adherence to the original Fels mission. Their stated aims were: (1) 'to secure sufficiently reliable criteria of fetal activity and reactivity so that these variables measured in the fetal period may be related to others measured in post-natal life' and (2) to clarify developmental sequences.

In a study of 30 pregnant women, observed repeatedly, Sontag and Richards (1938[300]) found marked variations in fetal heart rates within and between individuals. The mean rates tended to decrease slightly after the sixth month of pregnancy but rates exceeding 160 beats/minute occurred with a frequency of 6% in the last month of pregnancy indicating that caution should be exercised when rapid rates are used as an index of fetal distress. These findings were confirmed many years later by Welford *et al.* (1967). Sontag and Richards (1938[300]) also reported observations of 16 fetuses in a study that extended to 2 weeks after birth. The fetal and infant data for heart rates and for variability of heart rates in these Fels participants were not significantly correlated. These workers repeated the observations of Sontag and Wallace (1935b[306]) and showed that a vibrator placed on the maternal abdomen tended to increase the fetal heart rate, particularly late in pregnancy, but the effects they noted were not statistically significant. They reported that increases in fetal heart rates were associated with fetal movements but not with maternal heart rates, or maternal sleeping or eating.

Sontag and Newbery (1940[291], 1941[292]) reported the incidence and nature of fetal arrhythmias, which they considered were usually of sinus origin and influenced by external factors such as pressure and vibration. In group data, Sontag and Richards (1938[300]) found that increases in heart rate were accompanied by an increased number of extrasystoles.

This work on fetal heart rate was theoretically interesting and it provided some direction to future research, but it was limited by its restriction to brief time intervals.

### *Fetal movements*

The first report from the Fels Research Institute described an apparatus for recording fetal movements (Sontag & Wallace, 1933[303]). Four air-filled rubber sacs held firmly against the maternal abdomen simultaneously recorded the movements of all parts of the abdomen. Pens mounted on these sacs produced records on moving paper. Fetal movements caused differential shifts of the pens, whereas maternal breathing caused similar shifts in all the pens. This apparatus was used for only a brief period because it was cumbersome and close correspondence was shown between the records of fetal movements from the apparatus and those from maternal reports (Sontag & Wallace, 1935b[306]).

Instances in which women stated that fetal movements were increased

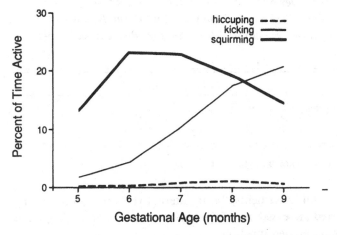

Fig. 3.1 The percentages of time spent hiccuping, kicking and squirming by fetuses at various gestational ages (data from Newbery, 1941).

with fear, anxiety, fatigue and auditory stimulation were reported (Sontag & Wallace, 1934[304]; Sontag, 1944a[279]). Richards, Newbery and Fallgatter (1938) considered that errors in the maternal reports of fetal movement may have obscured some real relationships. There was some consistency across time in the amounts of movement recorded for fetuses with month-to-month rank order correlations of about +0.7. There was not, however, any evidence of a diurnal rhythm or a relationship to maternal activity. Nevertheless, changes in the maternal basal metabolic rate during 2–6-month intervals were significantly correlated with the frequency of fetal movements, but not their intensity. After about 32 weeks of gestation, there was a marked increase in fetal movements and these movements increased in response to a sound stimulus. This response might be an index of neuromuscular maturation (Sontag & Wallace, 1935b[306]; Bernard & Sontag, 1947).

The possibility that the frequency of fetal movements could be associated with the fat content of the infant at birth was explored by Sontag (1940[277]). He found a significant negative correlation between the amount of fetal movement during the last 8 weeks of pregnancy and weight/recumbent length[3] which he used as an index of fat stores.

In another study of 12 women, it was shown that the frequency of fetal movements accounted for 30–70% of the variance in the performance of the infant on the Gesell scale at 6 months (Richards & Newbery, 1938; Richards & Nelson, 1938). The authors postulated that this could reflect bias in the maternal reports if more alert and intelligent mothers were more

Table 3.2 *Percentage deviation of basal metabolic rates from the standards of Boothby, Berkson & Dunn (1936) in the ninth month of pregnancy and 1 year after delivery (from Sontag, Reynolds & Torbet, 1944[299])*

|                          | N   | Mean      | SD   |
|--------------------------|-----|-----------|------|
| Ninth month of pregnancy | 158 | +5.89[a]  | 8.5  |
| One year after delivery  | 115 | −8.33[b]  | 9.7  |
| Gain in pregnancy        | 142 | +14.27    | 10.8 |

The difference between a and b is significant ($p < 0.01$).

likely to report fetal movements. In an alternative explanation, Sontag (1941[278]) suggested increased use of the nervous system during the period of myelinization improved function.

Newbery (1941) directed attention to possible differences in the types of fetal movements. She wrote: 'Some fetuses are constantly turning and squirming. Others keep the same general position but kick and thrust with hands and feet. Some fetuses have hiccups almost every day, others not at all.' The recognition of kicking and squirming appeared to be highly reliable with coefficients of 0.8 between measurements made 2 weeks apart. The percentages of time spent by the fetuses in these activities are shown in Fig. 3.1. Some trends were evident despite the marked variability. With advancing gestational age, there was an almost linear increase in the mean percentage of time spent kicking and there was a slight decrease in the percentage of time spent squirming after a gestational age of 6 months. Newbery suggested that these trends may be related to changes in the size of the fetus and the force it can exert or they may be responses of the fetus to slight anoxemia late in pregnancy.

In the Fels study, fetal hiccups were described by a mother as 'sudden, quick jerks or thumps, recurring at regular intervals every 2 to 4 minutes' (Norman, 1942). Fetal hiccups have a pronounced sound component and become less frequent minute by minute. Norman first noted them as being reported by mothers in the fifth month of pregnancy; they became more common as pregnancy advanced except for a decrease during the final 2 weeks. The hiccup scores were markedly variable, but did not differ with parity or obesity (Newbery, 1941). The incidence of fetal hiccups seemed to be related to changes in maternal position that may cause pressure on the fetus or the cord. The general conclusion was that the unreliability of the data did not justify further studies. Fetal hiccups are now recognized as a real phenomenon, but they are not known to be functionally significant.

Fels was unique among the US longitudinal studies in the emphasis placed on the prenatal period. The Fels scientists conducted many prenatal studies that required massive efforts and many resources; at least 15 000 hours of fetal movements were recorded. These studies tended to become institutionalized and, although subjected to some critical appraisal, this did not include consideration of their termination until 1970. Many recent changes have occurred in this scientific area resulting from the introduction of expensive equipment which almost mandates that studies be hospital-based (Cronk, 1983).

### Basal metabolic rate

Basal metabolic rate was a major component of the early Fels observations, as it was in the Denver Growth Study. Measures of basal metabolic rate are useful guides to the active cell mass of the body and to caloric needs. The Fels reports emphasize that the recorded values tended to be lower than the 'standards' because the latter had been obtained in clinics from patients who were anxious and were not accustomed to the procedure. Also, the Fels data were obtained in duplicate and the lower of the paired values was used in the analyses. Some of the Fels interest in basal metabolic rates concerned their possible associations with maternal and fetal tissue changes during pregnancy. Sontag, Reynolds and Torbet (1944[299]) reported basal metabolic rates from 158 Fels mothers who were free of pathological conditions that might have altered their basal metabolism. Despite marked variability, the values at the ninth month of pregnancy were significantly higher than those 1 year after delivery; the increases during pregnancy were negatively correlated ($r = -0.6$) with the values 1 year post partum (Table 3.2).

A further analysis was made of relationships between basal metabolic rates and some other variables in five pregnant participants each of whom had multiple 3-day balance studies (Pyle, 1938). A rationale was not presented for the associations tested. There were significant positive correlations of gestational age with urinary creatine and basal metabolic rate, but the correlations between urinary nitrogen and hemoglobin were negative. When the stage of pregnancy was held constant, there was a negative correlation between basal metabolic rate and hemoglobin. The author realized that the sample size was too small for definitive conclusions. This exploratory work exemplifies the far-ranging interests of Fels scientists.

### Familial and genetic studies

Many analyses within the Fels Longitudinal Study have been based on familial associations and, therefore, reflect the combined effects of genetic and environmental factors. An overview of these studies will be presented in this chapter; they are considered in other chapters in relation to the determinants of particular traits.

Between 1971 and 1977 two types of genetic data were collected from the Fels participants and their blood relatives: (i) 24 red cell markers, 17 blood proteins or enzymes and saliva factors that are completely genetically determined, and (ii) traits with a large genetic component that are also influenced by the environment. The traits in the second group include hand and foot dominance, tongue gymnastics, shapes of epicanthic folds and auricles, ear wax consistency, color vision, iris pigmentation, hair and skin color, phenylthiocarbamide (PTC) tasting ability, and dermatoglyphic patterns. Additionally, deoxyribonucleic acid (DNA) has been extracted from white blood cells and stored for future use.

In general, genetic data remain fixed for an individual throughout life. Consequently, it is hoped they will amplify the biological description of the Fels participants and that associations will be demonstrated between these variables and patterns of growth and maturation. A proposal to the National Institutes of Health to study possible linkages between variations in DNA and those in growth patterns has been funded recently. There is enthusiasm for this investigation which may greatly increase knowledge of the genetic control of growth patterns. It is only in the Fels Longitudinal Study that data exist that can provide this knowledge.

### *Familial studies*

Many data collected from Fels participants who share family relationships have demonstrated the presence of genetic control over growth and maturation (Sontag & Garn, 1957[286]; Garn & Rohmann, 1966a[141]). In a series of studies related to dental maturation, it was shown that the degree of tooth calcification at an age and the timing of dental emergence were more highly correlated within monozygous twin pairs than within dizygous twin pairs, and that these maturational traits were significantly correlated between sibling pairs (Garn, Lewis & Polacheck, 1960c[123]; Garn, Lewis & Kerewsky, 1965a[107], 1965b[108], 1967c[113]). Additionally, evidence was presented that the order of dental calcification, tooth size, and the number and pattern of dental cusps were under genetic control (Garn et al., 1965a[107], 1965b[108], 1965c[118]; Garn, Lewis & Shoemaker, 1956a[124]; Garn, Lewis & Kerewsky, 1967b[112], 1967c[113], 1967d[114]; Garn, 1966a[66]). Furthermore, the X chromosome may be important in some of these aspects of dental development (Garn *et al.*, 1960a[74], 1965a[107], 1965b[108]). This evidence for genetic control over dental maturation is not surprising since dental maturation is affected little by levels of nutrition (Garn *et al.*, 1965b[108]). In related studies, the genetic control of Carabelli's polymorphism on the first maxillary molar was found to be independent of tooth size and cusp number (Garn *et al.*, 1966g[76]; Garn, 1966a[66]).

The skeleton is influenced by both nutritional and endocrine variations. Nevertheless, evidence has been reported from the Fels study of genetic involvement in rates of skeletal maturation, the sequence of onset of ossification within groups of bones and the rates of periosteal bone growth (Garn & Rohmann, 1960a[134], 1966b[142]; Sontag & Lipford, 1943[288]; Reynolds, 1943; Garn & Shamir, 1958[164]; Garn *et al.*, 1961b[144]). Furthermore, there were sibling concordances in the patterns of ossification, especially for missing ossification centers in the foot and for the presence of pseudoepiphyses in the hand (Garn, 1966a[66]; Garn & Rohmann, 1966a[141]). The presence of a bony spur crossing the vertebral artery, as it passes superior to the posterior arch of the atlas, was shown to be inherited in a fashion consistent with a Mendelian dominant trait (Selby, Garn & Kanareff, 1955). A spur was present in some children when it was absent in the parents, and spurs were more common in boys than girls.

Other aspects of genetic involvement with skeletal growth were shown by analyses of data for recumbent length and stature in which there appeared to be evidence of X-chromosome effects (Garn, 1966a[66]; Garn & Rohmann, 1962a[136], 1966b[142]), and by analyses of parent–child correlations for stature (Garn & Rohmann, 1966b[142]). Himes and colleagues (1981) explored this matter further. They calculated correlations between the average statures of the two parents (mid-parent stature) and the statures of sons and daughters at ages from 2 through 18 years. These coefficients were about 0.4 at ages up to 13 years for the offspring and then increased rapidly with age in boys but not girls.

Analyses based on sibling pairings led Reynolds (1951) to conclude there was considerable genetic involvement in the thickness of subcutaneous adipose tissue, but later analyses of Fels data failed to confirm this (Garn, 1961c[59]). Significant correlations were reported, however, between the chest breadths of parents and stature, mental test performance and the rates of skeletal maturation in the offspring (Garn *et al.*, 1960a[74]; Garn, 1962b[61]; Kagan & Garn, 1963). These familial correlations and those for age at menarche and Gesell test item achievement (Garn & Rohmann, 1966a[141]) could be due to genetic or environmental influences.

Dermatoglyphic records of the fingers and palms were obtained from

Fels participants older than 8 years, and from their relatives. Dermato-glyphic patterns form between the 7th and 24th weeks of fetal life; a critical period for their development occurs at about the 13th week. There is marked genetic control over the development of dermatoglyphic patterns but the mode of inheritance, while almost certainly polygenic, is not clearly established. The local factors involved in their development are poorly understood. Since, once formed, these patterns do not change, they reflect the genetic and local environmental factors operating early in development and, therefore, have great biological interest.

The distribution of dermatoglyphyic patterns on digits was described by Roche, Siervogel & Roche (1979a[268]). In the Fels population, arches were common on digits II and III, but rare on digit V. Radial loops and radial loop closures were more common on digit II than on the other digits and, on each digit except digit I, plain whorls were more common than double loop whorls.

Ridge counts on the radial and ulnar sides of the fingers for unrelated Fels participants were subjected to principal components analysis (Siervogel, Roche & Roche, 1978). Three components were obtained that were consistent across sex and hand and represented three regions: (i) digit I, (ii) digits II and III, and (iii) digits IV and V. These are probably distinct developmental fields that provide varying local environments which interact with specific gene sets that are involved in the formation of dermal ridges. Additionally, there were significant correlations between digits in radial and ulnar ridge counts and in the totals of these (Siervogel, Roche & Roche, 1979).

Siervogel and colleagues (1978, 1979) derived a radio–ulnar whorl ratio to measure the asymmetry of ridge counts for digits that have whorls. This index increased from digit II to digit V, indicating that asymmetry favored the radial ridge count in digit II and favored the ulnar side in digits IV and V. A developmental gradient between digital areas in combination with local environmental factors would account for these findings.

Data from a three-generation pedigree were reported in detail because an extremely rare variant, absence of triradius d, was noted in one Fels participant (Roche, Roche & Siervogel, 1979c[275]). Some relatives of this participant had other rare dermatoglyphic features, but not an absent triradius d. Therefore, a sporadic non-genetic cause could have been responsible. A continuum of variants at this location was described, with absence of the triradius as the extreme.

In an unusual and interesting study, Meier, Goodson & Roche (1987) analyzed dermatoglyphic data in relation to whether individuals pass through pubescence relatively early or relatively late. Groups of rapidly and slowly maturing males differed significantly in ridge counts and digital pattern intensities while the corresponding groups of females differed only in palmar pattern types. Since dermatoglyphic patterns are established early in fetal life, the data suggested that the tempos of fetal development and of pubescent development are related. In each sex, those who were late to reach pubescence tended to have more complex dermatoglyphic patterns, although the affected areas differed between the sexes. This work provided suggestive evidence of a possible link between the rates of fetal and pubescent development. It is one of many exciting possibilities presented by the Fels data.

The failure of teeth to develop was used also as an index of variation in the intrauterine environment during the first trimester of pregnancy (Garn, Lewis & Kerewsky, 1963b[104]). The prevalence of missing teeth was established and it was demonstrated that absence of a third molar tooth tended to be associated with the absence of other teeth or reduction in their size. These studies by Meier and by Garn represent attempts to understand more about prenatal characteristics and their relationships to traits observed after birth.

In an investigation that exemplifies the widespread interests of Fels investigators, especially during the early years of the study, Sontag and Allen (1947[282]) reported that children with pulmonary calcification were more likely to be sensitive to histoplasmin than to tuberculin and the timing of the development of sensitivity, in relation to the development of calcification, was closer for histoplasmin than for tuberculin. Lung calcification tended to be familial, but there was no familial tendency for the histoplasmin skin reactions.

### Genetic studies

There was an intentional over-enrollment of triplets into the Fels study because of interest in comparing individuals within these trios, especially when the trios included monozygous pairs. This interest led to a paper by Robinow (1943) concerning the diagnosis of zygosity based on concordance for traits inherited in a Mendelian fashion. His statistical approach was based on concordance–discordance ratios from studies of randomly selected twins. This approach led to estimates of the probability of monozygosity and of polyzygosity within sets of twins or triplets when genetic traits were concordant.

Some studies of triplets focused on phenotypic similarities. A set of identical triplets was the subject of three reports. Sontag and Nelson (1933[290]) described their physical and mental traits. Later, Sontag and Reynolds (1944[297]) described the onset of ossification which tended to be the same in all the triplets, except for the metatarsals. Reynolds and Schoen (1947) described the growth and maturation of these triplets, including the distribution of subcutaneous adipose tissue and muscle thicknesses. Despite considerable similarity, many minor differences were noted in all the variables examined.

Linkage studies attempt to establish the presence of statistically significant associations between genes or genetic markers. Falk *et al.* (1982) and Spence *et al.* (1984), using Fels data in combination with data from other studies, demonstrated significant linkages between some pairs of genetic markers and the absence of linkage for some other pairs. Other analyses at Fels have utilized the records of traits that have a large genetic component. Contrary to some claims in the literature, it was shown that iris pigmentation, which is genetically determined, was not associated with hearing thresholds at various frequencies in children (Roche *et al.*, 1983b[256]).

Segregation analysis of pedigree data from 1152 individuals in 120 families, including many Fels participants, was used to examine alternative genetic models that could explain the patterns of inheritance of the ability to taste phenylthiocarbamide (PTC); some find this very bitter but others find it tasteless (Olson *et al.*, 1989). These models, unlike the usual ones, allowed for the occurrence of taster matings with non-taster offspring, as occurs in the Fels and some other studies. The best fit to the data was obtained with a two-loci model in which one locus controlled PTC tasting and the other controlled more general tasting ability.

Skin color is determined partly by genes and partly by environmental influences. Skin color is usually graded by its ability to reflect light. A series of interesting analyses concerned possible endocrine effects on skin reflectance measured at various body sites in the Fels participants (Garn, Selby & Crawford, 1956b[161], 1956c[162]; Garn & French, 1963[81]). The observations were made during the bright months, May through October, at sites exposed to a lot of solar radiation, those exposed to little radiation and those, such as the scrotum and the areola of the breast, where reflectance is markedly responsive to sex hormones. Despite the low coefficients of reliability for measurements 6 months apart, some clear differences and trends were established. In males, but not in females, reflectance decreased with age at the sites exposed to little solar radiation, but there were decreases at the areolar site in each sex with age, particularly after 50 years in the women. The marked decreases in areolar reflectance during pregnancy persisted through the first year after delivery and showed a significant tendency to accumulate with successive pregnancies.

The genetic studies made in the past at Fels, and those likely to be made in the near future, concern isolated topics that are not easily grouped as a set of hypotheses that could be supported by external funding. In the past, the Fels Fund might have helped, but that is no longer possible. Given the absence of graduate students, many of these possibilities are likely to remain unexplored in the forseeable future. Exceptions are (i) the set of familial and genetic hypotheses that is being addressed as part of the body composition studies (Chapter 7), (ii) the investigation of the genetic control of growth patterns that has just started and (iii) the genetic control of changes in serum lipid levels. Plans are being developed for studies of the last-mentioned topic.

# 4 Physical growth

'Growth is the only evidence of life.'

*John Henry Cardinal Newman* (1801–1890)

This chapter could be disproportionately long because so much has been achieved at Fels in the area of physical growth. To reduce its length, the growth of bones and teeth and studies of body tissues are described in Chapters 6 and 7 respectively. The present chapter will describe research concerning (i) the development and standardization of anthropometric methods, (ii) age changes in anthropometric variables, (iii) methods of growth assessment, (iv) secular changes in anthropometric variables, (v) determinants of growth, (vi) the final phase of growth, (vii) the prediction of adult stature, (viii) associations between growth and behavioral variables, and (ix) future directions. Fels research related to these topics that extends beyond the period of adolescent growth will be included.

## Development and standardization of anthropometric methods

The Fels Research Institute has long been recognized as a center of excellence in anthropometry. Few of the methods used at Fels are novel but they are described in considerable detail in the research protocol and they are applied with unusually high reliability. In discussing the assessment of nutritional status, Garn (1962c[62]) complained that anthropometry was often regarded as a set of crude procedures and little or no attention was given to the need for standardization and training. To meet the need for better standardization of anthropometric techniques between studies, a North American Consensus Conference was held in 1985 under the leadership of Tim Lohman. Fels staff members contributed significantly to the success of this conference, at which agreement was reached on the procedures for measuring 40 body dimensions (Lohman, Roche & Martorell, 1988).

Robinow (1968) considered the applicability of anthropometric procedures to field studies. He stated that crown–rump length was not useful during infancy although it can provide data about leg length as a percentage of total body length (relative leg length). This ratio could be important because its rate of increase is reduced by malnutrition during infancy. Despite this potential importance, crown–rump length was not reproducible in his data and was considered to be affected by the size and firmness of the buttocks. Robinow further recommended that choices be made between arm circumference and calf circumference and between biceps and triceps skinfolds because these measures were highly correlated.

Robinow (1968) considered the interpretation of head circumference in the context of nutritional assessment. He demonstrated positive associations between head circumference and body size (recumbent length and weight) in normal and in malnourished children.

Accuracy of measurement has long been recognized as important. As the Bible puts it: 'A false balance is an abomination to the Lord; but a just weight is his delight' (Proverbs 11:1). The potential introduction of automated equipment to increase accuracy was suggested several decades ago (Garn, 1962c[62]; Garn & Helmrich, 1967[89]). Automated equipment was described that can transfer anthropometric data directly to computers, thus eliminating errors during recording, transcribing and keyboard data entry, and that can be used to measure lengths and produce punch cards automatically.

A new inexpensive stadiometer, known as the Accustat® Ross Stadiometer, was tested by comparison with the expensive Holtain instrument and with Healthometer scales that have a vertical rod attached for the measurement of stature (Roche, Guo & Baumgartner, 1988b[239]). Repeated measurements of children with all three instruments showed the Accustat was the most reliable. Consequently, it was recommended for general use. A new sliding caliper (Mediform®) of moderate cost was shown to be slightly more accurate for the measurement of body lengths than much more expensive calipers that are in common use for research (Chumlea, 1985a[5]). The new caliper was recommended for the measurement of arm length, knee height, and similar body dimensions.

Comparisons were made among two expensive commercially available calipers and an inexpensive caliper developed by Ross Laboratories for the measurement of knee height (Cochram & Baumgartner, 1990). Intracaliper reliability was similar for all three instruments.

Many aspects of anthropometry, particularly the measurement of adipose tissue, knee height and somatotypes, were considered in recent critical reviews (Roche, Baumgartner & Guo, 1991a[213]; Chumlea, in press[13]).

## Age changes in anthropometric variables

It is commonly stated, as a truism, that human beings cannot be studied throughout their life spans because the investigator will not live long enough. This attitude ignores the possibility that a succession of investigators, using the same procedures, can conduct such studies. Figure 4.1 shows serial status-at-age values and annual increments for stature and recumbent length from 1 to 56 years in one Fels participant, reported by Roche *et al.* (1981a[235]). This paper was entitled 'The first seriatim study of human growth and middle aging' so that it would be a companion piece to Scammon's classic paper of 1927 entitled 'The first seriatim study of human growth.' The 1981 report contains what is believed to be the longest record of human growth ever made. In this case, five successive investigators were responsible for the data collection. The early part of Fig. 4.1 presents data for recumbent length and the later part presents data for stature. Both are measures of the total length of an individual. While these values are highly correlated, recumbent length is systematically greater than stature (Roche & Davila, 1974a[223]). The figure shows rapid, but decelerating, growth during infancy, a pubescent spurt at about 12 years, and only small increases in stature after 16 years that are followed by decreases after about 36 years. While this figure shows values for only one Fels participant, it illustrates the unusual duration of the Fels study. The Fels study is also unusual in its collection of growth data after 18 years. This has led to several interesting analyses that are described later in this chapter and were summarized by Roche (1975[182]).

Falkner (1971a) emphasized the changes in growth rates with age and the need to collect data that would allow the description of growth patterns. Later, summarizing for a WHO Expert Committee, Falkner (1977a) directed attention to the rapid changes in size during pubescence and the myriad other changes, such as alterations in mental and psychosocial development, that occur about this time. He pointed out that these changes have their origins long before puberty. Chumlea (1982[2]) also reviewed physical growth during adolescence. He directed attention to the hormonal control of the initiation of pubescence and described changes that occur during the maturation of secondary sex characters, including familial associations in their timing. Chumlea emphasized the variability in timing of sexual maturation and noted that this is more marked than the differences in the sequence of changes. Despite the variation in timing

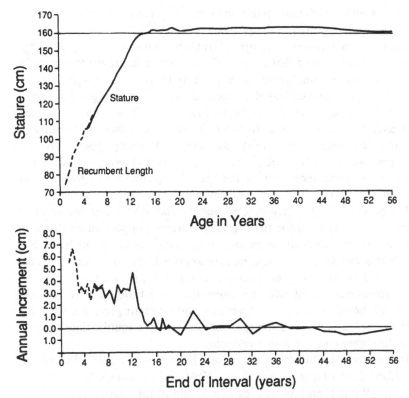

Fig. 4.1 Serial recumbent lengths and statures in one Fels participant from 1 to 56 years, with the annual increments. Note the pubescent spurt with its peak at 12 years and the decrease in stature after 36 years (data from Roche *et al.*, 1981a[235]).

between individuals, there are high correlations within individuals for the timing of grades and, consequently, the timing of later stages can be inferred from status at younger ages. He noted the similarity between the procedures for grading sexual maturity by direct visual examination and from photographs, and he criticized the theory that menarche occurs when a critical level of body weight or of total body fat is reached.

As Chumlea (1982[2]) noted, the skeletal changes during pubescence are generally larger in boys than in girls; this sex-associated difference is marked for shoulder breadth. Additionally, the later pubescent spurt in boys allows growth to continue about 2 years longer in boys than in girls. Viewed somewhat simplistically, this delay in boys is responsible for their greater adult stature and their longer legs and arms relative to stature. The timing of pubertal spurts tends to differ between bones with disto-proximal gradients within each limb: pubescent changes occur earlier in the bones of the hand and foot than in the proximal bones of these limbs.

Scalp thicknesses were investigated by Garn, Selby and Young (1954[163]). These workers showed that scalp thicknesses on radiographs tended to be greater in males than females after 10 years but not at younger ages. In each sex, the means increased to about 20 years but changed little from then to 40 years. Later, Young (1959), using similar data, reported increases in these thicknesses from 1 to 9 months and then little change to 3 years, after which there was a steady increase. At young ages, the greatest thickness was near the root of the nose but after 10 years the scalp was thickest on the posterior part of the head. Scalp thicknesses were greater for men than for women at corresponding sites.

Roche (1985b[200]) reviewed continuities and discontinuities in child growth, directing attention to the greatly increased opportunities that now exist to answer important questions using computer-based methods. For example, serial data for a variable can be analyzed by fitting a mathematical model to the data. Analyses of serial data for multiple variables depend on the implementation of two successive steps: (i) fitting a mathematical model to the serial data for each variable and estimating its parameters for each individual, and (ii) using these parameters in a multivariate analysis as if they were cross-sectional variables.

In two major studies, Kouchi and her co-workers (1985a, 1985b) described the patterns of growth in weight and recumbent length during infancy. Weight and weight gain are important, particularly during infancy, because of their associations with disease and mortality. Recumbent length is important also because slow growth in length indicates the presence of an influence that is retarding this physiological process. Additionally, recumbent length values assist the interpretation of weight, particularly as part of the ratio weight/recumbent length$^2$.

The data were from 441 Fels participants, including 87 parent–offspring pairs and 282 sibling pairs, all of whom had been examined at ages from near birth through 24 months with few missing data. A flexible mathematical model with three parameters (intercept, intrinsic rate of growth and growth pattern) was fitted to the serial data for each participant. The three biologically meaningful parameters summarized the growth of each infant in three values (coefficients) for weight and three values for recumbent length. The model fitted very closely to the observed data, even for participants with extreme coefficient values: the mean square residuals were less than the measurement errors. The fits for weight tended to be better in the girls than in the boys, but the fits to the recumbent length data were equally good in each sex.

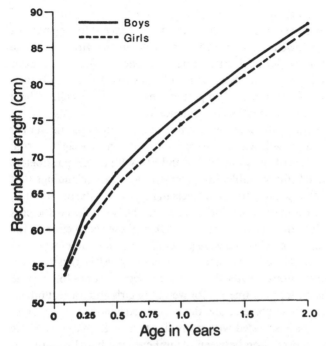

Fig. 4.2 Plots of recumbent length during infancy for boys and girls based on the means of parameters of mathematical models. (From Kouchi, M., Roche, A.F. & Mukherjee, D. (1985b). Growth in recumbent length during infancy with relationships to adult status and familial associations of the estimated parameters. *Human Biology*, 57, 449–72. Redrawn with permission from *Human Biology*.)

For both weight and recumbent length, the intercepts were significantly larger for the boys than the girls, showing that the boys had larger birth weights and larger values for recumbent length at 1 month. Similarly, the boys had significantly more rapid intrinsic rates of growth in weight and in recumbent length. The third parameter (growth pattern) was significantly smaller for the boys than the girls, indicating that the growth patterns were more curved in the boys than in the girls for these variables. Figure 4.2 shows plots of recumbent length for boys and girls based on the sex-specific means of the parameters of the model. This figure presents the mean patterns of growth and not the means of values at a series of ages considered cross-sectionally. The latter would be misleading. Because the boys had a more curved growth pattern, the curves for the two sexes were closer to each other at each end of the age range than near the center of this range.

Weight-at-an-age tends to be positively skewed, with a more marked tendency to large values than to small ones. Despite this, the parameters of the model for weight were symmetrically distributed except for the pattern parameter which showed slight positive skewness in the girls and some bimodality in each sex (Kouchi, Mukherjee & Roche, 1985a). The latter finding is potentially important because it indicates that there may be two distinct groups of infants in regard to patterns of growth in weight. These groups may differ environmentally or genetically.

In both boys and girls, the confidence limits of the pattern parameter for weight showed that growth was a square root function of weight during infancy. The estimates of the growth rate and growth pattern parameters for weight were markedly variable, in agreement with earlier findings that there are marked changes in percentile levels in individuals during infancy.

For each measure and each sex, the correlations between the coefficients were corrected for the error variance. After these corrections, the correlations among the coefficients were generally low and, therefore, most of the coefficients could be interpreted independently. After these corrections, the only correlations greater than 0.3 were negative ones for the rate of growth with the pattern parameters for weight in girls and for recumbent length in each sex; these correlations demonstrated tendencies for rapid intrinsic growth to be associated with more linear growth patterns. There was also a negative correlation between recumbent length at 1 month and the rate of growth in recumbent length during infancy, which showed that catch-up growth in recumbent length during infancy should not be assumed.

Correlations were calculated between the parameters of growth during infancy and measurements of the same individuals at 18 and 30 years (Kouchi *et al.*, 1985a, 1985b). Many of these correlations were significant but most were modest, ranging from 0.2 to 0.6 for weight and from 0.2 to 0.5 for recumbent length. The parameters for growth in infancy did not explain more than 30% of the variation in the adulthood measures of weight, stature, head circumference and calf adipose tissue thicknesses. These correlations showed little consistency between the sexes or between weight and recumbent length, but significant correlations were more common for the intercepts and the rates of growth than for growth patterns.

Birth weights (intercepts) were positively correlated with weight and weight/stature$^2$ at 18 and 30 years. Recumbent length at 1 month in females was significantly correlated with weight at 18 and 30 years. Additionally, recumbent length at 1 month and the rate of growth in recumbent length were significantly correlated with stature at 18 and 30 years in each sex. Head circumference at 18 and 30 years was significantly

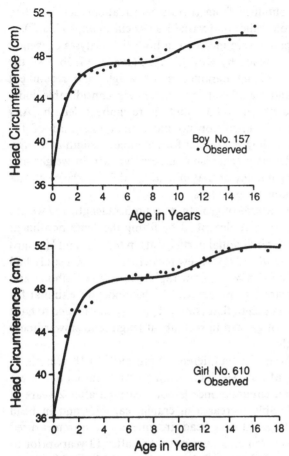

Fig. 4.3 The observed values for head circumference and the fitted models for a boy (No. 157) with a very low value for final head circumference, and for a girl (No. 610) with a very low value for initial size. (Roche, A.F., Mukherjee, D. & Guo, S. (1986b). Head circumference growth patterns: Birth to 18 years. *Human Biology*, **58**, 893–906[257]. Redrawn with permission from *Human Biology*.)

correlated with the rate of growth in weight during infancy. The low associations between growth rates for weight during infancy and adult status for adipose tissue thicknesses are consistent with other Fels findings obtained using different analytic procedures (Cronk *et al.*, 1982a; Roche *et al.*, 1982a[264]). They provide further evidence that fatness during infancy is not significantly associated with the level of fatness in adulthood.

It is particularly interesting that the correlations between the rates of growth in weight during infancy and adult stature were as high as those between stature at 2 years and adult stature. These findings are in general agreement with those obtained from regression equations developed to predict adult stature from childhood variables (Roche *et al.*, 1975a[272]).

Many of the participants were included in both the analyses of weight and of recumbent length made by Kouchi *et al.* (1985a, 1985b). All the correlations between matching parameters for weight and recumbent length were positive and significant, but there was considerable independence between growth in weight and in recumbent length. Any parameter for weight accounted for no more than one-third of the variation in the corresponding parameter for recumbent length at an age. This indicates that different mechanisms underlie growth in weight and recumbent length during infancy or that individuals differ in the responses of these processes to influencing factors.

In another study, the patterns of growth in recumbent length and weight from 3 months to 6 years were described by fitting the Jenss non-linear model to the serial data for individual participants in the Fels and Harvard Growth studies (Byard *et al.*, 1991). A few participants in each study had average or low values in infancy but grew rapidly in early childhood and had high values at 18 years. The parameters of the model were similar for the Fels and Harvard data except that Harvard participants tended to have less curvilinear patterns of growth in recumbent length and slower linear rates of growth in weight.

Patterns of growth in head circumference from birth to 18 years have been described (Roche, Mukherjee & Guo, 1986b[257]). Unusual data were available; typically, head circumference is not measured after 2 years of age despite the considerable increases in cranial capacity and in head circumference after that age. At Fels, head circumference is now measured at all ages but these measurements were not made after 13 years prior to 1952 and they were not made after 2.5 years from 1952 to 1969. Head circumference is an important index of brain size, particularly during infancy when the cranium is thin. Although the changes in head circumference are small after 13 years, head circumference differs markedly between individuals and it is a potentially important predictor of fat-free mass (Chapter 7).

A mathematical model was fitted to serial data for each individual, using the non-linear least squares procedure (Roche *et al.*, 1986b[257]). This model had four parameters: (i) the estimated head circumference at 18 years, (ii) a dimensionless parameter that was a function of size at birth, age at peak growth rate during pubescence, and the velocity at peak growth rate, (iii) the rate of change in acceleration (cm/year$^{-1}$), and (iv) a critical age when both the slope (cm/year$^{-1}$) and the acceleration (cm/year$^{-1}$) were zero. This final parameter indicated the onset of the pubescent spurt. The model fitted well and therefore the four parameters effectively summarized the serial data, even for participants with unusual values as illustrated in Fig. 4.3.

The means for head circumference in the boys were greater than those for the girls at birth and at 18 years, with differences of 0.5 and 1.5 cm respectively. Furthermore, the boys were about 9 months older than the girls at the onset of the pubescent spurt in head circumference and about 18 months older when the maximum rate of growth occurred. Growth in head circumference during pubescence was more rapid in the girls than in the boys, and pubescent growth was slower in each sex when pubescence was delayed.

Growth patterns in the Fels, Denver, and Harvard Growth studies and in the Guidance Study at Berkeley were compared (Thissen *et al.*, 1976). Because some of the studies included numerous familial relationships, the samples were limited to the oldest child of each sex within each family. The double logistic model fitted well, although it tended to overestimate the data in mid-childhood and after puberty and to underestimate the data in early adolescence. To overcome problems due to systematic differences between the studies in the timing of the scheduled examinations, a least squares fitting procedure was used in which each observation was weighted by the inverse of the number of observations during each year.

The variances of the parameter estimates were similar for all the studies but some subtle differences in the growth patterns were demonstrated by sex-study interactions. Additionally, there were significant study effects in the maximum velocities of the prepubertal and adolescent components of the model and in adult stature. The general conclusion from this comparison was one of similarities rather than differences between patterns of growth in these studies although, in each sex, mature stature was about 2 cm less in the Harvard study than in the other studies.

The mean differences between the growth patterns of girls in the Fels and Berkeley Guidance studies are shown in Fig. 4.4. These groups of girls had almost identical mean adult statures although the Fels girls had about 3 cm less prepubertal growth. These small differences in growth patterns between the four studies suggested there were only minor regional variations in growth for US white children and that results from studies of Fels participants were likely to be applicable to other groups of children. In passing, it should be noted that growth patterns for stature are more regular for girls than for boys in the Fels study (Roche *et al.*, 1975a[272]).

Fels data for weight and recumbent length have been used to provide reference data for body surface area from birth to 3 years (Roche, Guo & Moore, 1991[244]). Published equations were used to calculate empirical

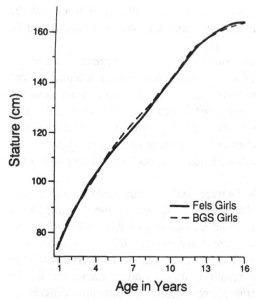

Fig. 4.4 Median patterns of growth in stature for females in the Fels and Berkeley Growth Studies. (Thissen, D., Bock, R.D., Wainer, H. & Roche, A.F. (1976). Individual growth in stature: Comparison of four U.S. growth studies. *Annals of Human Biology*, 3, 527–42. Redrawn with permission from *Annals of Human Biology*.)

values for selected percentiles and these were smoothed across age by nonparametric kernel regression. The percentiles increased rapidly from birth to about 6 months, after which the increases were slower. At all ages, the values for boys tended to be larger than those for girls. Sex-specific reference data are needed from 6 months to 2 years. These data can be used to judge the normality of body surface area which is important in the standardization of some physiological and metabolic variables such as renal function and caloric needs.

### Methods of growth assessment

If measurements of growth-related variables are made at only one age, the values can be regarded as increments since conception. These values are useful in screening, and in the evaluation of infants and children. Repeated measurements can be plotted on status reference charts and their positions and changes in position can be judged subjectively, relative to the age changes in cross-sectional data that are displayed in the reference charts. This approach can show whether the increments between examinations are smaller or larger than average, but evaluations made in this way are likely to be inaccurate soon after birth and during pubescence when growth rates change rapidly. Reference data for increments are needed to judge whether the variation from the central tendency is unusual. The importance of increments in growth assessment, particularly during infancy, was stressed by Garn (1962a[60]). If increments are evaluated, unusual rates of change will be recognized earlier than if judgments are based on serial data for status (Roche *et al.*, 1989a[242]; Guo *et al.*, 1988).

### Reference data

Fundamentally, growth assessment depends on comparisons between recorded data and reference data. The information needed for comparisons in growth assessment should be referred to as 'reference data.' This term may not have originated at Fels but the Fels staff have certainly contributed to its widespread use (Hamill *et al.*, 1977, 1979). The distinction between reference data and 'standards' is important. The word 'standard' is appropriate in the physical sciences. For example, until 1960, the standard meter was the length of a particular metal rod kept in Paris. We do not have a standard child or a 'correct' (optimal) value for the weight of a 5-year-old boy. Healthy boys of this age can differ in weight by large amounts and without significantly increased risks of disease. We can, however, obtain percentiles for weight in 5-year-old boys by the study of large samples and judge the weights of other 5-year-old boys by reference to these percentiles.

In an approach that was innovative at the time, standard deviations were used to develop reference data arranged as T-scores and as standard deviation levels for several variables (Sontag & Reynolds, 1945[298]; Reynolds & Sontag, 1945). These 'composite sheets' allowed the simultaneous assessment of weight, stature, number of ossification centers and illness. These charts were used routinely to monitor the growth progress of Fels participants until 1970, and they were used by many outside the institute.

Garn (1965[65]) considered the applicability of growth reference data derived from North American children to children in developing countries. He postulated that a few well-known sets of reference data would be preferable to a large number of sets for use in world-wide nutrition surveys. This is similar to the current WHO view that a single set of reference data facilitates international comparisons. This WHO view is appropriate for the major body dimensions but not for some other characteristics, e.g., fat patterns. It is based, in part, on knowledge that the distributions of major growth-related variables would be similar in many countries if environmental differences were removed. Some differences would remain, however, particularly for populations that differ in body build (e.g., weight–

stature relationships, relative leg length). Garn (1965[65]) made the interesting suggestion that US reference data for the growth of children of short parents be used tentatively in countries where the populations have not been studied, and he suggested that privileged groups in a developing country, that are genetically the same as the general population, could indicate the expectations for growth under better environmental conditions.

Later, Garn (1966b[67]) cogently argued in favor of population-specific sets of growth reference data. In this publication, he suggested that these reference data be based on mid-parent stature but pointed out that this would require an immense study, even if restricted to a few categories of parental stature. The need for large samples of parents of different statures has been circumvented in analyses at Fels by treating the data in a continuous fashion, as will be described later (Himes *et al.*, 1981).

Falkner (1971a) addressed general aspects of growth assessment. He outlined the purposes of such assessments and stressed the 'whole child' and the recognition of retarding environmental influences. He claimed that 'norms' of size at an age, or a series of ages, were of little use to those wishing to screen children. In this context, he used 'norm' as synonymous with 'mean.' He stated that the normal ranges should be provided although these ranges are statistically based and do not lead to easy evaluation of an individual child as normal or abnormal.

Later, writing on behalf of a Committee of the International Union of Nutritional Sciences, Falkner (1972) recommended the establishment of sets of cross-sectional growth reference data for each country that should be derived from subgroups of national populations for which growth was not retarded by the environment or disease. As Falkner noted, in many countries, it is difficult to find large groups of children who meet this criterion. As he pointed out, national reference data must be constructed with careful attention to study design, sample size, and quality control. Futhermore, appropriate statistical procedures must be applied to obtain and smooth reference percentile values. Even then, limitations will remain due to the genetic diversity present within most countries. This diversity is important if the genetic differences cause growth variations, but a considerable literature indicates that genetic differences cause only a minor part of the growth variations between populations. This implies that a single set of reference data could be used internationally if it were obtained by excellent procedures from a population free of retarding influences.

Falkner (1978a) noted that the reference data from developed countries for growth during infancy are largely derived from formula-fed infants. This led him to conclude that breast-fed infants at the lower percentile levels of these reference data may not need supplementation. In this paper, he directed attention to the need to group low-birth-weight infants on the basis of gestational age and he stressed the importance of increments in the assessment of growth, especially during infancy.

Roche and Falkner (1975[232]) critically reviewed the reference data available for US children and expressed concern about their applicability. Reluctantly, they recommended the Kaiser/Permanente charts for infants and the Iowa charts for older children. At about the same time, a conference report concluded that, despite recognized differences in growth among ethnic groups in the US, a single national set of reference data would be unlikely to cause serious errors (Roche & McKigney, 1975[253]). This recommendation, together with others, soon led to the publication of the first set of US national growth data. These data are displayed in the National Center for Health Statistics growth charts for weight, recumbent length, stature, weight-for-recumbent length, weight-for-stature, and head circumference (Hamill *et al.*, 1977, 1979). The data for ages from birth to 3 years came from the Fels Longitudinal Study, and those for ages 2 to 18 years came from US National Surveys. Originally it was expected that these reference data would be used widely in the US. It was not foreseen that they would be used in other countries and that WHO would recommend their world-wide use to assist international comparisons.

The measurement of total body length is necessarily made with the subjects recumbent during infancy, but standing erect when they are older. The Hamill reference data are for recumbent length to 3 years and for stature after 2 years. Roche and Davila (1974a[223]) showed these measures were almost exactly correlated, but recumbent length is the larger. The differences between these measures were the same for both sexes except from 11 to 14 years when they were significantly larger in girls. The mean difference was about 1.2 cm but this value differed between studies, presumably due to variations in methods of measurement.

Other sets of reference data are needed to assess the growth of special groups of children or for uncommon measurements. It may be difficult, however, to locate suitable data. Consequently, Roche and Malina (1983[252]) compiled a two-volume *Manual of Physical Status and Performance in Childhood* to meet the need for a convenient source of reference data. This manual contains tabular data for a wide range of variables relating to body size and proportions, maturity, performance, function, body composition, and hormonal, dietary, physiological, and biochemical status. To the best of their ability, the authors included data from all relevant studies in North America published between 1941 and 1981; many of these utilized Fels data.

Several reviews from Fels concern growth assessment in normal and abnormal children. Methods for the assessment of growth and maturity and for the study of determinants of growth and maturation were reviewed in a book resulting from a NATO Advanced Study Institute (Johnston, Roche & Susanne, 1980). Later, Chumlea (1986a[8], in press[13]) reviewed methods for the assessment of growth in children with developmental disabilities. He noted the scarcity of reference data for particular groups and the need to measure many of these children in unusual positions. When handicapped children are measured standing, their posture should be examined carefully to be sure the data will be valid. The use of arm span as an alternative to stature was discussed but emphasis was placed on the use of recumbent anthropometric procedures for children who are unable to assume the correct positioning for the measurement of stature.

Roche (1978a[184], 1978b[185]) described methods of growth assessment that were disseminated widely in a booklet entitled *Pediatric Anthropometry*, jointly authored by Moore (Moore & Roche, 1982, 1983, 1987). This material has had a major influence on clinical anthropometry in relation to children. It was stressed that growth data are sensitive indicators of health and development and that serial growth data, either plotted on status charts or used to calculate increments, can help monitor the progress of a child during nutritional intervention or the treatment of a disease. The recommended procedures for obtaining common growth measurements were described in detail. It was stressed that the data should be plotted and also recorded in tabular form; the latter allows checks as to whether the plotting was accurate and it is from the tabular data that increments must be calculated if there is concern about the growth of a child.

Increments can provide more sensitive information about the normality of growth progress than status at an age or at a series of ages. These increments should be calculated for 1-month intervals during the year after birth and for 6-month intervals at older ages. Tables of consecutively numbered days were provided to assist the calculation of increments when, as is common, the interval between measurements is not exactly 1 month or 6 months (Ross Laboratories, 1981).

Guidelines were given by Moore and Roche (1982, 1983, 1987) for the interpretation of growth data by comparison with reference values for status or increments. Since stature tends to be similar within families, observed statures should be adjusted for the average of the statures of the two parents (Roche, 1978a[184], 1978b[185]; Himes *et al.*, 1981). Both the observed and the adjusted statures should be plotted. This approach, which was made widely available by Ross Laboratories (1983), allows the statures of children to be compared with the usual reference data after removing the effects of the parental stature on the stature of the child. This may greatly alter the interpretation of the child's stature and thereby influence clinical management. The work done at Fels that provided a full range of parent-specific adjustments for childhood statures is described later in this chapter.

The level of maturity is clearly important for the assessment of growth early in infancy when prematurity has a large effect on growth status and growth rates. The growth of premature infants can best be evaluated by comparison with reference data that are specific for premature infants. An alternative is to 'adjust' the chronological age of the premature infant by subtracting the number of weeks by which the infant is premature from the chronological age and thus obtain a 'conception-corrected' age, as noted by Falkner (1977b). This approach assumes that there are no systematic differences in growth between premature and full-term infants after this adjustment has been made and that the growth of a premature infant soon after birth is the same as if the pregnancy had continued to term. The need for age-adjustment due to prematurity decreases with age; this reduced need is met, at least to some extent, by the deceleration in the rate of growth after birth. This deceleration reduces the effect of the age-adjustment at older ages during infancy.

Maturity status is also important in relation to the timing of pubescence. Rapidly maturing children tend to be taller and heavier than the general population before pubescence and they have early pubescent spurts. After pubescence, their size moves closer to the mean as they experience an early deceleration of growth. Opposite trends occur in slowly maturing children. Knowledge of the maturity status of a child, either from skeletal age or the development of secondary sex characters, can allow better judgments of the normality of growth.

As stated by Moore and Roche (1982, 1983, 1987), serial values of weight and stature plotted in relation to age should be regular relative to reference percentile lines. Irregularity can be due to errors of measurement or of plotting, particularly if the irregularity is restricted to one variable. When the most recent point is 'out-of-line,' another examination should be scheduled if the irregularity is not removed by re-measurement and re-plotting. A real change in growth status, relative to percentile lines, should cause concern if the change is marked, or occurs near the extreme percentiles.

Parenthetically, it could be noted that: 'It is better to be short and good,' as stated by Mozart. Because growth measurements are distributed in a continuous fashion, some children must be near the ends of the distributions. These children, whether small or large, are not necessarily abnormal. Nevertheless, there is reason for concern when growth, which is a physiological process, proceeds at an unusual rate. The many possible causes include diseases, abnormal dietary intakes, psychosocial deprivation, congenital syndromes and familial short stature. It should be stressed that growth can be normal in the presence of a disease and that growth can be abnormal in the absence of a disease (Moore & Roche, 1982, 1983, 1987).

The purposes of growth assessments in children with renal diseases have been reviewed (Potter *et al.*, 1978). Growth assessments of these children can assist better understanding of the course of these diseases and the effects of treatment. Furthermore, serial growth assessments may help identify the types of renal diseases that most affect growth. Recommendations have been made about the timing and frequency of serial growth assessments in relation to the timing of dialysis and of renal transplantation, and the additional data that should be recorded at the time of these assessments. The paper by Potter and his associates recommended methods by which growth records should be utilized and interpreted. These methods include adjustments to observed statures for parental stature, the calculation of increments and comparisons with reference data on the basis of both chronological age and skeletal age. The latter is particularly important in judging the likely effect of the disease on adult stature.

Multivariate statistical analyses of cross-sectional data are needed commonly to compare a study population with a reference population; often access to the raw data is needed for both groups. Other statistical methods are used to test particular types of hypotheses. For example, cluster analyses can determine whether a group can be divided into statistically separate subgroups, discriminant analyses can determine the importance of particular variables in distinguishing between subgroups of individuals, and multiple regression procedures can estimate the relative importance of particular variables in predicting a continuous outcome such as blood pressure.

The use of standard deviation scores for several variables, derived for each individual, was recommended for the analysis of group data in children with renal disease (Potter *et al.*, 1978). This approach was used more than 30 years earlier by Sontag and Reynolds (1945[298]) and by Reynolds and Sontag (1945). These scores can be treated statistically in the same way as any other set of numbers. They may be used to compare children with renal disease to normal children, to compare the growth of children with various types of renal disease, and to evaluate the effects of treatment. It was concluded that predictions of adult stature are unlikely to

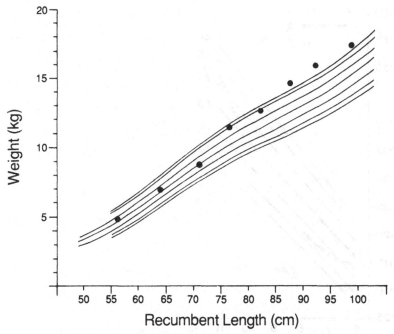

Fig. 4.5 Serial values for weight (kg) for one boy, plotted against recumbent length (cm) and related to national percentile levels (Hamill *et al.*, 1977).

be accurate in children with renal disease, except perhaps in those with uncomplicated renal tubular acidosis who, if treated adequately from a young age, may reach normal adult statures.

The assessment of growth in abnormal children was reviewed by Roche (1978a[184], 1978b[185]). He discussed the purposes of these assessments and directed specific attention to procedures for measuring recumbent length, stature, sitting height, weight, head circumference, upper arm circumference, and skinfold thicknesses, and for grading maturity. The necessary equipment was described and the sources from which it can be obtained were listed. The need for training and for documentation of reliability was emphasized, particularly for the assessment of skeletal age. References were provided to sets of recommended reference data.

In assessing the growth of abnormal children, as for normal children, it is important to interpret weight in relation to stature. Clearly, tall children tend to be heavier than short children. A convenient approach is to plot weight in relation to stature (weight-for-stature) on charts of reference data (Hamill *et al.*, 1977, 1979). Weight-for-recumbent length, of which an example from Fels is shown in Fig. 4.5, is nearly independent of age. The

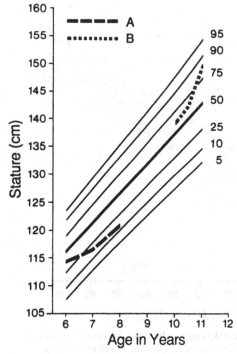

Fig. 4.6 Serial stature data for two boys plotted relative to reference percentile levels. See text for details.

interpretation of weight-for-recumbent length (weight-for-stature) depends on the relationship between trunk length and leg length. Children with relatively short legs are likely to have high values of weight-for-recumbent length. Usually these proportions are judged from the ratio of crown–rump length to recumbent length (sitting height to stature at older ages) because crown–rump length (sitting height) is a measure of trunk length. The growth of abnormal children can also be judged using model-fitting procedures (Holm *et al.*, 1979).

Roche (1978a[184], 1978b[185]) made suggestions regarding the interpretation of changes in 'channels.' In this context, 'channels' are the zones between adjacent selected percentile levels on growth charts. If serial statures are plotted against chronological age, a change of channel indicates a change in the rate of growth compared with the reference population. Thus Boy A in Fig. 4.6 had a slower rate of growth in stature than the reference population from 6 to 7 years and his stature percentile changed from the channel between the 25th and 50th percentiles to that between the 10th and 25th percentiles. Boy B in Fig. 4.6 changed channels between 10 and 12 years from that between the 50th and the 75th percentiles to that between the 75th and 90th percentiles. This represents more rapid growth than is usual in the reference data. Many increases in percentile level near the age of pubescence, as for Boy B in Fig. 4.6, reflect rapid maturation and early pubescence, while many decreases at about this age reflect slow maturation and late pubescence.

When there are changes in the rate of growth in stature, relative to reference data, it is important to determine whether there is an associated change in the rate of skeletal maturation. When increments in stature are plotted against skeletal age, the mean and median stature increments should differ little from those relative to chronological age but the variance will differ in a way that is not documented. This procedure is likely to be restricted to research studies; skeletal age assessments are time-consuming and standardization is difficult.

Plotting serial values for an individual on growth charts for status-at-an-age commonly gives a false impression of steady growth because the vertical scale is small (Falkner, 1977b). For many clinical and research purposes, it is better to calculate increments in growth and compare them with corresponding reference data. Falkner (1971a) noted that environmental influences had larger effects on weight increments than on stature increments and he noted the value of increments in the recognition of catch-up growth, particularly in low-birth-weight infants. Later, Falkner (1978a) directed attention to the large increments in head circumference in low-birthweight infants after birth. These increments are larger in premature infants, who reach normal values by 2 to 3 years of age, than in infants who are small for gestational age.

The need for incremental reference data was recognized early by the Fels staff; Robinow (1942a) published one of the earliest sets of such data. Robinow provided means and standard deviations of increments for the intervals from birth to 1 month and from 1 to 3 months, and then each 3 months to 1 year, and each 6 months to 6 years, with annual increments from birth to 6 years. Robinow reported that the distributions of increments were near normal. Nevertheless, almost all sets of incremental data are skewed and the presentation of percentiles would have been more appropriate (Garn & Rohmann, 1964b[140]). Robinow (1942a) did not adjust the data for variations in the lengths of the intervals and he combined data from the two sexes without testing the significance of the differences between them. Despite its importance, this early work received little clinical attention.

Falkner (1973a, 1978b) stressed the importance of reference data for increments and emphasized that they must be derived from accurate data (Sobel & Falkner, 1974; Falkner, 1978a). The attention given to detail during the recording of Fels data made them ideal for this purpose. Consequently, a set of reference data for increments during 6-month intervals was developed by Roche and Himes (1980[247]) who had a much larger data set than that available to Robinow. These reference data were originally presented in graphs but were later published in tabular form (Baumgartner *et al.*, 1986a). During the development of these reference data, adjustments were made for variations in the lengths of the intervals and it was shown that these increments did not demonstrate seasonal or familial effects. Earlier, however, Reynolds and Sontag (1944) reported slight seasonal trends in stature increments and larger ones in weight with peaks for stature increments in the summer and peaks for weight increments in the winter.

The reference data of Roche and Himes (1980[247]) were for recumbent length and head circumference (birth to 3 years), stature (2 to 18 years) and weight (birth to 18 years). For each of these variables, selected empirical percentile levels for each 6-month interval were smoothed across age using low-term Fourier transforms. Simple methods were provided by which users could adjust for variations in the intervals between measurements, together with guidelines to assist interpretation of the findings. In particular, it was stressed that serial percentile levels for increments within individuals were much less stable than corresponding values for status; the correlations between successive percentile levels for increments ranged from −0.4 to 0.7. Therefore, serial increments often showed changes in 'channel.' Reference data for increments allow earlier and more precise recognition of unusual growth than do reference data for status, but those presented by Roche and Himes (1980[247]) did not take into account the marked variations in the timing of the pubescent spurt. Therefore, while they were suitable for the assessment of groups of children unselected for age at pubescence, they were less satisfactory for the assessment of individual children as they pass through pubescence. This problem would be solved if there were an effective method of predicting the timing of pubescence from prepubertal data. In its absence, the charts that indicate the differences associated with rapid and slow maturation can only be applied retrospectively.

This work of Roche and Himes did not result in new insights but it provided graphs and tables, for use by clinicians and health-care providers, that were derived from data for a population similar in growth status to US national samples (Roche & Himes, 1980[247]; Baumgartner *et al.*, 1986a; Hamill *et al.*, 1977, 1979; Roche & Hamill, 1977[245], 1978[246]).

The growth of infants must be monitored at brief intervals. Therefore, reference data are needed for increments during 1-month intervals. In accordance with the accepted view, Falkner (1978a) claimed that measurements must be made at regular intervals to provide reference data for increments. Since there are few serial studies of large groups of infants in which measurements were made each month, an alternative approach was adopted to generate suitable reference data. Mathematical models fitted to sets of serial data for individuals allow interpolations to any ages within the range of the data. Of course, there must be a sufficient number of measured values spaced so that they can provide an adequate description of the growth changes, and the models must fit well.

Using Fels data recorded at seven ages from 1 through 24 months, a three-parameter mathematical model was fitted to serial values for weight, recumbent length and head circumference for each participant (Guo *et al.*, 1988; Roche et al., 1989a[242]). This model had the form:

$$f(t) = a + b\sqrt{t} + c \log t + e$$

where $f(t)$ is the measurement (cm, kg) at age $t$ (years); $a$, $b$ and $c$ are estimated parameters, and $e$ is an error term. Since the model fitted very well to the recorded data, interpolations were made to ages 1 month apart and 1-month increments were calculated. Empirical percentile levels were obtained for age intervals from 1 to 2 months through 11 to 12 months. In the publication containing these reference data, it was recommended that follow-up examinations be made of infants with values for weight, recumbent length, or head circumference outside the 5th to the 95th percentile range for status, and that their rates of growth be judged by comparing the observed increments with these reference data. These reference data for one-month increments demonstrated rapid decreases in the rate of growth during infancy (Guo *et al.*, 1988; Roche *et al.*, 1989a[242]). Estimates of 1-month increments could have been extended to older ages but the amounts of growth during 1-month intervals after 12 months were too small to allow the accurate categorization of children.

The use of reference data for 1-month increments in head circumference in the early recognition of abnormal growth is illustrated in Fig. 4.7a, which shows serial head circumferences for two theoretical girls each of whom had a head circumference at the 25th percentile at 1 month. It was assumed that Girl A grew slightly faster than the 95th percentile for 1-month increments and that Girl B grew slightly slower than the 5th percentile for these increments. When the data for these girls were plotted against the reference data for 1-month increments, the unusual growth of each girl was noted at 2 months (Fig. 4.7a). This would alert the clinician

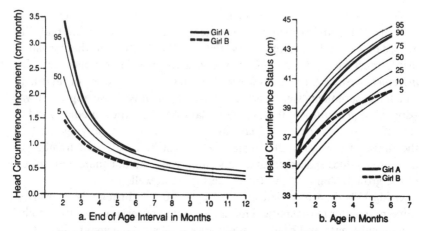

Fig. 4.7 Selected percentiles for 1-month increments and for status in head circumference for girls during infancy. In Fig. 4.7a, head circumference increments for two girls are plotted; the corresponding status values are plotted in Fig. 4.7b. See text for details.

who might re-examine these girls 1 month later or begin laboratory investigations immediately. In Fig. 4.7b, the serial data for these girls were plotted relative to the National Center for Health Statistics reference data for status that were derived from the Fels study. With this approach, there would be little clinical concern about Girl A but there might be concern about the declining growth status of Girl B by the age of 5 months.

These reference data for weight, recumbent length, and head circumference (Guo *et al.*, 1988; Roche *et al.*, 1989a[242]) are applicable to any 1-month interval from 1 to 12 months. For example, the interval could be 4 to 5 months or 4.2 to 5.2 months. Additionally, increments during intervals ranging from 3 to 5 weeks can be adjusted to approximate what would have been observed during a 1-month interval before they are compared with the reference data.

In an unusual but effective approach, growth data from treated phenylketonuric children were compared with Fels data (Holm *et al.*, 1979). Since all the data were serial, a mathematical model was fitted to the values for each child, after using a linear analysis of variance to estimate values at the scheduled ages. The model fitted well and showed remarkably close agreement between the groups in recumbent length, but the phenylketonuric children were lighter than Fels children from 2 to 3 years, and heavier at 4 years. The coefficients for head circumference did not differ between the Fels and phenylketonuric groups for boys, but in the girls the phenylketonuric group tended to have a higher value (intercept) at birth and more rapid deceleration. This work was an impressive demonstration of one way in which the Fels data can assist the interpretation of data from other studies.

### Secular changes in anthropometric variables

The term 'secular changes' refers to differences between individuals or groups that are correlated with year of birth but independent of age (Roche, 1979a[187]). For example, differences between the weights of 10-year-old girls born in 1900 and those born in 1970 would constitute secular changes if they were observed in the same population group. Secular changes in growth, by definition, reflect changes across time in environmental factors that influence growth. Therefore secular changes are relevant to studies of environmental effects on growth even if the specific aspects of the environment that are responsible remain unknown. Secular changes in growth have received considerable attention because of the striking acceleration of growth and maturation in Western Europe and North America during this century. One wonders whether Shakespeare had secular changes in mind when he wrote: 'Young boys and girls are level now with men.'

Secular changes in growth were the topic of a symposium organized by Roche (1979a[187]). In the introduction to this monograph, he stated: 'The study of secular trends is based, necessarily, on survey data from representative national populations or large well-defined groups within these populations: these groups must be measured or observed at several points in time, ideally separated by several decades or even longer.' Nevertheless, few studies of secular changes have been of defined groups that did not change due to migration. Much of the literature reviewed in this symposium was not based on Fels data, but knowledge of the extent of secular changes in the US, and in other countries, is important in relation to the general applicabililty of the findings derived from the Fels data. It is believed that secular increases in stature occurred in the US at least until 1965 but have now ceased or become much smaller. Chumlea (1982[2]) also reviewed this topic and concluded that the large increases in body size and rates of maturation may have ceased in some developed countries. In many other countries, secular changes have not occurred.

When only cross-sectional data are available and age is unknown, as occurs in excavated skeletal remains, it is difficult to estimate secular changes because age changes cannot be excluded. To assist the separation of secular changes from age changes, radiographic tibial lengths and statures were compared in a group of 288 adults that included many Fels participants (Hertzog, Garn & Hempy, 1969). Since the length of the tibia

Table 4.1. *The estimated and observed mean losses of stature (cm) due to aging (data from Hertzog et al., 1969)*

| Age (years) | Men estimated | Women estimated | observed |
|---|---|---|---|
| 35.0–44.9 | 0.95 | −0.06 | 0.24 |
| 45.0–54.9 | 0.97 | 0.01 | 0.80 |
| 55.0–64.9 | 1.93 | 2.93 | 2.40 |
| 65.0–74.9 | 2.19 | 6.15 | — |
| 75.0–87.0 | 3.12 | 6.47 | — |

does not change with age during adulthood, differences in tibial length, relative to year of birth, must represent secular changes. Differences in stature adjusted for tibial length, within age groups, were then used to estimate the loss of stature in individuals with aging.

In each sex, the tibial lengths were greater for those born between 1900 and 1920 than for those born before or after this. Estimated decreases in stature due to aging showed a loss in men from early adulthood onwards but losses in women occurred only after 55 years when the decreases were much greater than in men (Table 4.1). The validity of these conclusions is dependent on the absence of a secular trend in the relationship of tibial length to stature. Hertzog indirectly documented such an absence by showing a close correspondence between his estimates of statural loss and the observed losses in 52 women. Additionally, Himes (1979), after a comprehensive review of the literature, concluded that there has not been a secular change in the ratio sitting height/stature. This secular constancy of body proportions may occur for tibial length/stature relationships also.

Secular changes for weight have been slight in the Fels study. For example, Garn and French (1967[82]) reported only slight tendencies for boys to become heavier and girls to become lighter as groups born in progressively later years were considered. Using data within age groups, Byard and Roche (1984) reported regressions of recumbent length and stature on dates of examination, which are directly related to dates of birth. There was evidence of a significant linear trend for girls during childhood and significant small increases in the early years of the Fels study that were reversed later. Additionally, head circumference spurts during pubescence now occur earlier than in the past and the maximum rate of growth during the spurt has increased (Roche *et al.*, 1986b[257]). Furthermore, there have been only small secular trends for weight/stature$^2$ in the Fels data (Siervogel *et al.*, 1989a).

The small secular changes in the Fels population are in contrast with some other reports. Bock and his colleagues, after the analysis of data for three generations of Fels families, reported that, within families, later-born children tend to be taller than others of the same sex (Bock, Sykes & Roche, 1985; Sykes, 1985; Bock & Sykes, 1989). Additionally, they showed that, when the Fels participants and their relatives were placed in one of three groups by date of birth, the mean adult statures within families increased in each sex from the first to the second group but not from the second to the third group. These intergenerational findings within families indicated that the secular increases at Fels ceased about 1950.

Other analyses of two-generation families showed differences between parents and offspring measured at corresponding ages during growth (Bock & Sykes, 1989). In the males, the mean parent–offspring difference was almost constant from 1 to 18 years. In the females, the differences at 1 year and at 18 years were similar but differences were absent or slight during pubescence. The increase of about 3–4 cm in stature between parents and adult offspring of the same sex was not influenced by decreases in parental statures due to aging. These parent–offspring differences remained about constant for offspring born between 1932 and 1968, indicating that the rate of secular increase within families did not decrease during this period (Fig. 4.8). The contrast between the continuing secular increase in stature within families and the lack of such an increase in the total Fels data set was largely due to the recruitment of new families after 1940 in which the adult statures were less than those for the families recruited earlier.

The analyses by Kouchi and her associates (1985a, 1985b) of patterns of change in weight and recumbent length during infancy included participants who were born between 1929 and 1970. Possible secular changes in infant growth were examined by placing the participants in four groups on the basis of year of birth. Each of these groups included about one decade of birth years and was identified by the mean birth year (1935, 1946, 1958 and 1969). The differences between the groups were small but some were significant. Birth weight and recumbent length at one month were smaller in the 1946 group than in the 1934 group in each sex, indicating a slight negative secular trend. These negative secular trends in the early years of the Fels study could reflect the early enrollment of many participants with parents who were members of the Faculty of Antioch College in Yellow Springs. Additionally, in the boys but not the girls, birth weight and recumbent length at 1 month increased significantly from 1946 to 1968 and the patterns of growth in weight became significantly more curvilinear from 1946 to 1958. The changes from one decade to another in birth weight

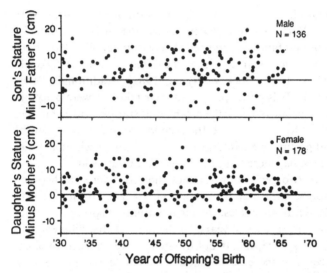

Fig. 4.8 Plots of the differences between the adult statures of parents and offspring of the same sex, plotted against the year of birth of the offspring. (Bock, R.D. & Sykes, R.C. (1989). Evidence for continuing secular increase in height within families in the United States. *American Journal of Human Biology*, 1, 143–8, copyright 1989. Redrawn with permission from Wiley-Liss, Inc.)

and in recumbent length at 1 month for boys are shown in Fig. 4.9. While the changes were somewhat parallel between weight and recumbent length, they were larger for birth weight than for recumbent length.

In another approach to the analysis of possible secular changes, the parameters of the individual curves fitted to weight and recumbent length were regressed on year of birth and also on (year of birth)$^2$ to estimate linear and non-linear associations. These analyses showed significant linear associations with birth weight and the rate of growth in recumbent length, tendencies to more linear growth patterns in weight for each sex, and increases in the rate of growth in weight for girls. There were some significant negative correlations between (year of birth)$^2$ and the parameters of the model that showed the direction of the changes altered about 1952. Prior to that year, birth weight was decreasing, the patterns of growth in weight were becoming more curvilinear and the rate of growth in recumbent length was decreasing; these tendencies were reversed after 1952. There were many siblings in the sample, but birth order was not associated with these secular changes.

Roche *et al.* (1986b[257]) examined possible secular changes in the patterns of growth in head circumference. The participants were placed in three groups, on the basis of date of birth. The onset of the pubescent spurt in

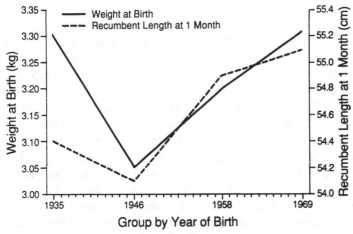

Fig. 4.9 Mean values for weight at birth and for recumbent length at 1 month, estimated from mathematical functions for boys grouped by year of birth. (Kouchi, M., Mukherjee, D. & Roche, A.F. (1985b). Curve fitting for growth in weight during infancy with relationships to adult status, and familial associations of the estimated parameters. *Human Biology*, **57**, 245–65, and Kouchi, M., Roche, A.F. & Mukherjee, D. (1985b). Growth in recumbent length during infancy with relationships to adult status and familial associations of the estimated parameters. *Human Biology*, **57**, 449–72. Redrawn with permission from *Human Biology*.)

head circumference occurred later during the early years of the Fels study compared with data recorded more recently. Additionally, the maximum rate of growth during the spurt was larger in those born recently. These secular changes in growth patterns were small and they were not associated with differences between birth groups in head circumference at birth or at 18 years.

### Determinants of growth

Because the participants in the Fels study are, with few exceptions, healthy and well nourished, the Fels data are not ideal for the study of factors that affect growth. Nevertheless, some interesting observations have been made.

#### *Type of infant feeding*

In the Fels population, there were only small differences in growth between breast-fed and formula-fed infants from birth to 3 months, but at 6 months, a larger percentage of the breast-fed infants were below the fifth percentile in weight and recumbent length (Roche, Guo & Moore, 1989c[243]).

### Fatness

Several analyses of Fels data have concerned associations between fatness and measures of growth and maturity. Some of these have involved long bone lengths, muscle thicknesses and ages at onset of ossification, but the account in this chapter will be restricted, in the main, to reported associations between fatness and either recumbent length or stature. Garn and Haskell (1960[88]) reported that correlations between adipose tissue thickness over the tenth rib and stature were higher in girls than boys at most ages. In the boys, these correlations were in the range 0.2–0.4 until 13.5 years, after which they were near zero. In the girls, the corresponding correlations were about 0.4 until 14.5 years, after which they were about 0.2 until 17.5 years when they were near zero. While most of these correlations were significant, it is clear that obesity explains only a small part of the variance in stature. Garn and Haskell concluded there was a linear relationship between stature and adipose tissue thickness over the tenth rib, within the limits of the study.

Garn (1962b[61]) reported correlations of about 0.3 between maternal weight in the first trimester of pregnancy or maternal weight gain during pregnancy, with birth weight. Although these correlations were significant, the maternal characteristics explained only a small part of the variance in birth weight. There were even lower correlations between maternal size and recumbent length at birth, which may have been due partly to the difficulties inherent in the measurement of recumbent length at this age. Sibling correlations in the Fels data for birth weight, recumbent length, and segment lengths at birth were about 0.3 (Garn, 1961c[59]).

Low correlations (about 0.2 to 0.3) between radiographic adipose tissue thicknesses and recumbent length in newly born and older infants were reported from early Fels studies (Garn, 1956a[46], 1958a[52]). At 1 month, infants above the median for fatness had larger recumbent lengths than those below the median. Nevertheless, the increments in recumbent length after 1 month were essentially the same in the two groups of infants and the differences were slight or non-existent after 6 months. Garn ascribed the marked differences at 1 month to variations in birth weight between the groups, but did not comment on possible group differences in recumbent length at birth. In another study, however, Garn (1962b[61]) did not find an association between fatness and recumbent length during infancy. He concluded that an asymptote had been reached beyond which the further accumulation of fat did not accelerate growth.

After analyzing data from children aged 7.5 and 11.5 years, Reynolds and Asakawa (1948) reported high correlations (r = 0.7 to 0.8) between weight and stature. More direct associations between fatness and stature were presented by Reynolds (1951) in his landmark monograph. Reynolds reported correlations between radiographic adipose tissue thicknesses at six sites and stature at 7.5, 11.5 and 15.5 years (Table 4.2). The correlations tended to be highest for thicknesses at the calf and trochanteric sites. The correlations with stature were relatively high for boys aged 7.5 years and they decreased with age in each sex except from 7.5 to 11.5 years in the girls. Reynolds also reported partial correlations between the sums of adipose tissue thicknesses at these six sites and stature, with weight held constant. These coefficients were negative at each of the three ages in each sex and ranged from $-0.3$ to $-0.7$.

Garn and Haskell (1960[88]) concluded that there was a linear relationship between adipose tissue thickness at the tenth rib site and stature. They fitted a linear function to within-age plots of normalized adipose tissue thickness against stature and, of course, this did not reveal any within-age curvilinear relationship or asymptote. Almost all the correlations were significant to 12 years but later they decreased, particularly in the boys, and almost all were non-significant at older ages. The correlation coefficients tended to be higher for the girls than for the boys, but, even in the girls, fatness accounted for only a small part of the variance in stature.

It is interesting to note that the Fels data have been analyzed in relation to the same topic across a span of 26 years. Answers are never complete, but the data base gradually enlarges and better statistical methods become available. Himes and Roche (1986) returned to this topic when they analyzed the relationships of radiographic adipose tissue thicknesses at the medial calf and tenth rib sites to stature. Their findings were not in agreement with the earlier conclusions by Garn and Haskell (1960[88]) and by Garn (1962b[61]) that after infancy body size and maturational status can be increased almost indefinitely by stepping up the caloric surplus.

In the study by Himes and Roche, adipose tissue thicknesses were adjusted for radiographic enlargement and for the small differences between the scheduled and the actual ages at examinations. Linear and non-linear relationships (cubic polynomials) were evaluated. In the boys, the correlations for the calf increased from near zero in infancy until 11 years, after which they decreased. These correlations for the calf and for the tenth rib site in boys were significant at most ages from 5 to 15 years. In the girls, the correlations for the calf were significant at most ages from 3 through 12 years, and those for adipose tissue thicknesses at the tenth rib site were significant at all ages from 0.5 through 14 years.

In boys, the non-linear model had significantly larger coefficients than the linear model for the calf thicknesses at 9 and 11 years and the tenth rib thicknesses at 9 through 11 years. In the girls, the non-linear model had

Table 4.2. *Correlations between stature and radiographic adipose tissue thickness at six sites (data from Reynolds, 1951)*

| Sites | Boys | | | Girls | | |
|---|---|---|---|---|---|---|
| | 7.5 years | 11.5 years | 15.5 years | 7.5 years | 11.5 years | 15.5 years |
| Calf | 0.52[a] | 0.23 | 0.07 | 0.34[b] | 0.32[b] | 0.15 |
| Trochanter | 0.31[b] | 0.20 | −0.07 | 0.15 | 0.36[b] | −0.10 |
| Waist | 0.47[a] | 0.14 | −0.02 | 0.21 | 0.24 | 0.02 |
| Chest | 0.27 | 0.19 | −0.02 | 0.20 | 0.20 | −0.08 |
| Forearm | 0.27 | 0.16 | −0.18 | 0.20 | 0.14 | 0.02 |
| Deltoid | 0.39[a] | 0.21 | −0.05 | 0.16 | 0.17 | 0.04 |

[a] $p < 0.01$.
[b] $0 < 0.05$.

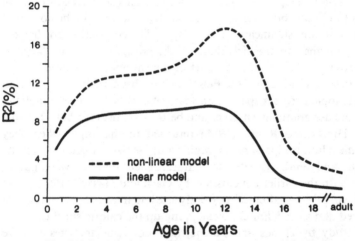

Fig. 4.10 Smoothed percentages of variance accounted for by linear and non-linear relationships among adipose tissue thickness at the 10th rib site and stature in girls. (Himes, J.H. & Roche, A.F. (1986). Subcutaneous fatness and stature: Relationships from infancy to adulthood. *Human Biology*, **58**, 737–50. Redrawn with permission from *Human Biology*.)

significantly larger correlations than the linear model for the calf thicknesses at 3, 11 and 12 years and for the tenth rib thicknesses at 9 through 14 years (Fig. 4.10). The non-linear correlations, but not the linear ones, increased markedly during pubescence. Plots of adipose tissue thicknesses against stature showed that increases in thickness beyond a 'threshold' level were not associated with further increases in stature (Fig. 4.11).

### Familial studies

As noted earlier, there are significant correlations between the statures of parents and their children. These should be taken into account when childhood stature is evaluated. Garn (1961c[59]) categorized parents as tall, medium, or short and compared the statures of the children from various types of parental pairings, e.g., tall–tall, tall–medium. He concluded that 'height–weight tables that ignore parental size are unrealistic. They apply only to foundlings....' Later, they published means at various ages for stature relative to the average of the statures of both parents (mid-parent stature) (Garn, 1966b[67]; Garn & Rohmann, 1966a[141]). These tables did not distinguish between recumbent length and stature. To assist the application of these data, Garn and Rohmann (1967[143]) reported numerical estimates of mid-parent statures when the mother's stature was known but the father's was only categorized to tall, medium or short. Since the distributions of childhood statures within mid-parent stature groups were not provided, these data are not suitable for clinical application.

Garn (1958a[52]) reported a low correlation (r = 0.3) between the statures of fathers and the recumbent lengths of infants at birth. Later, in a slightly different approach, Garn (1962b[61]) showed marked differences between the statures of offspring from tall × tall matings and those from short × short matings. This led him to conclude that adult stature was not controlled by genes that were purely cumulative. Garn showed that serial statures and serial lengths of a hand bone (second metacarpal) were more alike for monozygotes than for dizygotes, but even the monozygotes showed some differences. He concluded that the controlling factors involved in these differences affected the whole organism.

In 1981, Himes and his co-workers published a detailed monograph dealing with parent–child similarities in stature and provided a method to adjust recorded childhood statures for mid-parent statures. Parent–child correlations for stature may be affected by a tendency for spouses to be similar in stature. There was some evidence of such mating in the Fels study where the husband–wife correlation for stature was 0.2, but this did not affect the adjustments calculated by Himes *et al.* (1981) because they were derived from mid-parent–child regressions. Parent–child correlations could be affected also by marked differences in the environments during growth for parents and children, but this possibility can be disregarded for the Fels data because secular changes have been small (Garn & French, 1967[82]; Roche *et al.*, 1975a[272]). The estimates made by Himes *et al.* (1981) could have been affected by the inclusion of several siblings of the same sex in some families, but it was shown that restriction of the data base to one child of each sex within each family did not alter the results.

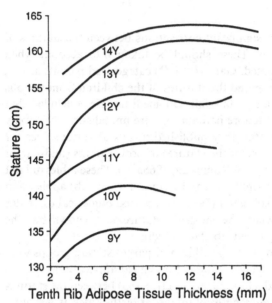

Fig. 4.11 Polynomial curves of stature against adipose tissue thickness at the 10th rib site in girls aged 9–14 years. (Himes, J.H. & Roche, A.F. (1986). Subcutaneous fatness and stature: Relationships from infancy to adulthood. *Human Biology*, **58**, 737–50, Redrawn with permission from *Human Biology*.)

Himes *et al.* (1981) analyzed data from 271 parents and 586 children. The data for children were adjusted to ages at scheduled examinations and the parental statures were those recorded closest to 30 years. They used a linear model of the form:

*child stature* = $a + b$ (*mid-parent stature*) + *error term.*

The coefficients for the regressions between mid-parent statures and the statures of the offspring, which are the b terms in the equations, increased rapidly from birth to 2 years and then more slowly to 7 years (Himes *et al.*, 1981). After 7 years, the coefficients increased in a near linear fashion in each sex, with a considerably steeper slope for the boys than for the girls. These authors tested other models that allowed for non-linear relationships, differential effects of the statures of the mother and father, and interaction effects of the parental statures. These alternative models did not reduce the standard errors in either sex except for slight reductions with the interactive model at some ages in the boys but not in the girls (Himes, 1984a).

The regression coefficients and the standard errors were smoothed across age and were adjusted relative to the National Center for Health

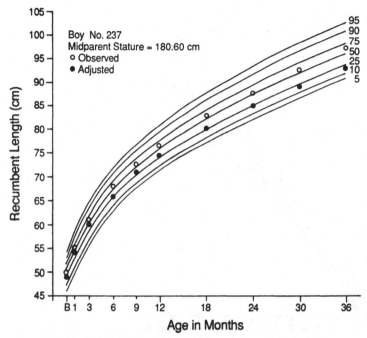

Fig. 4.12 Observed and adjusted recumbent lengths for mid-parent stature in a Fels boy with tall parents shown relative to NCHS percentiles.

Statistics reference data. The final parent-specific adjustments to observed childhood statures were presented for age- and sex-groups within categories of recumbent length and stature and distributed in a clinically useful format by Ross Laboratories.

These adjustments to the observed recumbent lengths and statures of children provide a powerful tool for distinguishing between children with 'normal' genetic causes of unusual stature and those with diseases that affect growth. For example, use of the adjustments will show whether a child with a stature less than the 5th percentile who has short parents is within the normal range, given the statures of the parents. These adjustments should be applied when the stature of a child is unusual and the parental statures are also unusual. Himes (1984b), after a thorough literature review, concluded that the parent-specific adjustments that had been derived from US whites were equally applicable to US blacks.

The measured and the adjusted values should be plotted on the National Center for Health Statistics growth charts (Hamill *et al.*, 1977, 1979). Figure 4.12 shows pairs of measured and adjusted recumbent lengths for an infant with tall parents. The measured values are near the median level

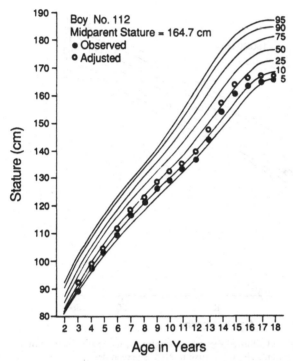

Fig. 4.13 Observed and adjusted statures for mid-parent stature in a Fels boy with short parents.

but the adjusted values are near the 25th percentile. Figure 4.13 shows corresponding data for an older boy with short parents. The observed values are near the 10th percentile but most of the adjusted values are near the 25th percentile. In these children, the adjustments led to considerable changes in percentile levels.

These adjustments are important in the clinical assessment of children in whom growth in stature is unusual. The adjustment makes important differences if the average stature of the parents is unusual; in these circumstances, it allows the child to be compared indirectly with the 'population' to which he or she belongs. This adjustment for inherited factors makes it easier to evaluate the cause of an unusual stature and choose the best management for it.

Familial correlations for recumbent length and stature were reported by Byard, Siervogel and Roche (1983) using data for pairs of related participants measured at corresponding ages. These correlations tended to decrease during pubescence, but were mainly influenced by the degree of relationship, age and sex. The sibling correlations were consistently higher than the parent–offspring correlations except after 15 years when both sets were similar. This method of analysis did not allow a separation of genetic, environmental and generational effects.

Later, Byard, Siervogel and Roche (1988) applied new path analysis techniques to annual measurements of recumbent length and stature to assess the transmissible and non-transmissible components of family resemblance for stature. The proportion that is transmitted reflects the joint influences of genes and culture, whereas the non-transmissible component is environmentally determined. The data they analyzed were for parents and their offspring all measured at corresponding ages. These workers calculated correlations using maximum likelihood procedures with adjustments for the variations among families in the number of children. Many of the sibling correlations exceeded 0.5, indicating that factors other than genes contributed to family resemblance. The importance of cultural inheritance from the parent was indicated by the finding that the sibling correlations were uniformly higher than the parent–offspring correlations. At about 4 years, the transmissible and non-transmissible components had their maximum coefficients. After this age, the coefficients for the non-transmissible components decreased rapidly, but those for the transmissible component were fairly constant.

Garn (1961c[59]) reported that parents with broad chests tended to have tall and heavy children with increased values for weight relative to stature. These effects were more marked in boys than girls and indicated a familial influence on this aspect of frame size.

Intrafamilial correlations were calculated for the three parameters of the model that was fitted to serial data for weight and recumbent length by Kouchi *et al.* (1985a, 1985b). All the correlations were positive but the parent–offspring correlations were not significant. The low parent–offspring correlations for birth weight were consistent with other evidence that birth weight is mainly determined by non-genetic maternal factors. All the sibling correlations were significant except those for some growth pattern parameters. The patterns of the correlations did not suggest sex-linked inheritance for changes in weight and recumbent length during infancy.

On average, parent–offspring and the sibling pairs have 50% of their genes in common. The lower correlations for parent–offspring pairs than for sibling pairs may have reflected differential environmental effects. Siblings share a common intra-uterine environment which is not the case for parent–offspring pairs. Also the postnatal environment is likely to be more similar for sibling pairs than for parent–offspring pairs. Much earlier,

Sontag and Garn (1957[286]) described four siblings in whom, from birth to 10 years, there were similar patterns of rapid growth in stature and weight, rapid maturation and slow motor development. Their findings indicated that these age-related changes may be genetically determined.

Correlations were calculated between values derived from the model fitted to serial data for head circumference by Roche *et al.* (1986b[257]) for parent–offspring pairs and for sibling pairs. These analyses took advantage of the unusual nature of the data set which contained measurements of these related participants at matching ages. All the sibling correlations, but none of the parent–offspring correlations, were significant. Even the sibling correlations were only low to moderate indicating that there was only slight genetic regulation of growth in head circumference. The differences in the correlations between parent–offspring and sibling pairs presumably reflected greater similarity of the environment (diet, illness) for siblings measured a few years apart than for parents and their offspring who were measured several decades apart. These findings matched those from other Fels studies of growth patterns in recumbent length, stature and weight (Byard *et al.*, 1983; Kouchi *et al.*, 1985a, 1985b).

Garn (1958a[52]) reported low correlations ($\sim 0.3$) between siblings for birthweight. Additionally, Garn (1961c[59]) noted the occurrence of some sibships in which there was a clear pattern of decreases with age in standard scores for stature.

### The final phase of growth

Fels scientists have conducted important analyses related to the final phase of growth in stature and the changes that occur in weight during the same period. The Fels data are ideal for such analyses. In contrast, many other growth studies have been based on school populations in which there has been considerable attrition at older ages and data collection has ceased at 18 years.

Garn *et al.* (1961b[144]) reported that the total increment in stature after the completion of epiphyseal fusion in the hand–wrist varied from zero to 4 cm, being larger in girls than boys. In a subsequent paper, Garn and Wagner (1969[170]) reported that slow-maturing children can grow as much as 7 cm after 17 years, although the mean total increases were 2.3 cm for males and 1.2 cm for females. Some of this variation could be due to differences in the rates of maturation between the long bones and the bones of the hand–wrist.

In later work, data from 194 Fels participants who had been examined to at least 28 years were used to determine the age at which adult stature was reached (Roche & Davila, 1972[222], 1975[225]; Roche, 1976[183]). Two

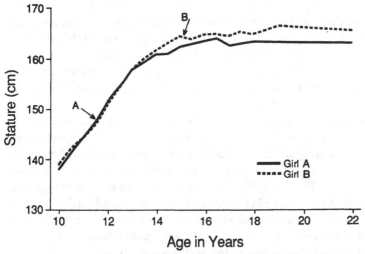

Fig. 4.14 Serial stature data for two Fels participants who differed markedly in age at menarche (see text for details).

mathematical models were fitted to the serial data recorded after the peak rate of growth in stature during pubescence (PHV). A polynomial (non-linear model) was fitted to the earlier data and a horizontal line was fitted to the later data. Piecewise regressions were calculated in which progressively more points were included in the 'earlier data.' The goodness of fit, measured by the differences between observed and predicted points for the two functions combined was calculated for each junction between the two models. The junction at which the goodness of fit was maximal was accepted as the age at which adult stature was reached. This method was effective in all the participants except a few in whom there were only small changes in the goodness of fit between successive junctions.

It was estimated that adult stature was reached at median ages of 21.2 years for males and 17.3 years for females. The amounts of growth after 16 years were considerably larger for males (2.4 cm) than for females (0.9 cm), and there were corresponding differences between the sexes in the amounts of growth after 18 years. Additionally, it was shown that stature increased about 1.5 cm in males and about 1.0 cm in females after the bones of the leg were completely mature (Roche & Davila, 1972[222], 1974b[224], 1975[225]). This growth must occur in the trunk.

In the females, growth in stature continued for almost 5 years after menarche with an increase in stature of 7.9 cm during this time. The increments in stature after menarche were negatively associated ($r = -0.5$) with the age of menarche, showing that late-maturing girls tended to have less growth in stature after menarche. This is illustrated by the data for a late- and an early-maturing girl in Fig. 4.14. Girl A, who reached menarche at 11.6 years, grew much more after menarche than Girl B, who reached menarche at 15.2 years. This relationship is represented also in Fig. 4.15 which shows selected percentiles of the total growth in stature after peak height velocity and menarche in relation to the ages at which these events occurred (Roche, 1989a[207]).

Later analyses using data from 520 Fels participants with serial data to at least 20 years showed the total increments in stature after peak height velocity were slightly larger in males (17.8 cm) than in females (15.8 cm). As expected, the annual increments decreased markedly as the interval from PHV increased (Roche, 1989a[207]). There was almost as much growth during the first year after PHV as in the next 4 years combined. There was a similar decrease in the annual increments in stature after menarche: the increments during the first year after menarche were greater than the sums of the annual increments during the next 4 years.

Changes in weight after PHV were analyzed using data from 229 Fels participants with serial data extending to at least 22 years (Roche & Davila, 1974b[224]; Roche et al., 1975c[227]). Piecewise regression was applied, as was done for stature, but in this case a polynomial was fitted to the earlier data and a linear function, that was not necessarily horizontal, was fitted to the later data. The junction at which the goodness of fit was maximal for the two functions combined was accepted as the age at which there was a change in the pattern of growth in weight. In some participants, more commonly than for stature, there were only slight differences in the goodness of fit between successive junctions and it was not possible to determine precisely the age at which the change in the pattern of growth occurred. The median ages for these junctions were 21.2 years in the males and 18.2 years in the females, with large ranges in each sex. For example, in males the 10th and 90th percentile ages were 16.9 and 25.1 years respectively. These ages were positively correlated with the weights at the junctions in the males (r = 0.4) but not in the females, and with age at PHV in each sex (males 0.4; females 0.3).

The total increments in weight were considerably larger in the males than in the females after particular ages and developmental events, such as PHV, and the completion of maturation of the bones of the leg. Furthermore, the functions fitted to the later data had much steeper slopes in the males than in the females, showing that the males gained weight more rapidly than the females after the change in the pattern of growth occurred. As was observed for stature, the increments in weight after menarche were negatively correlated (r = −0.4) with age at menarche.

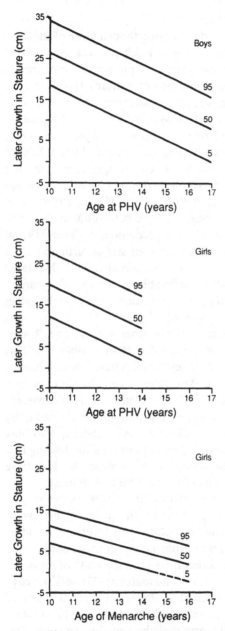

Fig. 4.15 Selected percentiles for the total growth in stature after peak height
velocity and menarche in relation to the ages at which these events occurred.
(Roche, A.F. (1989a). The final phase of growth in stature. *Growth, Genetics
and Hormones*, 5, 4–6[207]. Reproduced with permission from McGraw-Hill, Inc.)

### The prediction of adult stature

In an early study, Garn (1966b[67]) used the notion that children of the same age have achieved similar percentages of their adult statures. He reported 'multipliers' by which adult stature could be predicted from present stature for children with 'average' mid-parent statures (167–175 cm). Neither the sample sizes nor the errors of the estimates were given, but it was claimed that, for children with skeletal ages within 0.5 sd of the mean, the accuracy was as good as that of more complex approaches. Garn recognized the importance of skeletal age and recommended that it be used in place of chronological age. In this paper, Garn argued that the knee skeletal age should, in principle, be better than hand–wrist skeletal age because regions of the body differ in maturity levels (Garn, Silverman & Rohmann, 1964e[167]) and the bones near the knee contribute markedly to stature attainment. In a review of stature prediction, Falkner (1971a) directed attention to the low correlations between size at birth and adult size and the rapid increase in these correlations soon after birth. He stated that adult stature became reasonably predictable by the age of 3 years.

A new method for the prediction of adult stature was published in monograph form by Roche and his co-workers (1975a[272]); this work was summarized and made more generally useful in other publications (Roche, Wainer & Thissen, 1974a[270], 1974b[271], 1975b[273]; Roche, 1980b[191]). This investigation was undertaken because many families have marked interest in the adult statures of their children. This interest can change to serious concern if the present statures of the children are unusual and the parents consider it likely that their adult statures will be unusual also. This family concern increases the psychological burdens of such children, who may have social difficulties. In this context, a physical examination leading to a diagnosis may be indicated, but many such children do not have pathological conditions; they are part of the normal distribution.

In children with diseases that alter growth in stature, accurate predictions of adult stature would help selection for possible therapy to increase the potential for growth and the results of such therapy could be monitored by serial predictions. This approach is better than assessing the elongation and maturation of the skeleton because the units of measurement and the distributions differ for length and maturity. This difference in units makes it difficult to judge differential effects on these two sets of variables. Predictions of adult stature could be used also to evaluate therapeutic effects for children with hormonal abnormalities or congenital heart disease and to assist the evaluation of 'catch-up growth' whether this follows medical, dietary or other types of intervention. Despite these claims for the utility of adult stature predictions, the predicted values are not guarantees. They are only estimates for an individual and these estimates have errors that are often expressed as confidence limits.

A comprehensive review of the literature showed that predictions should be derived from a combination of variables that, almost certainly, would include the present age, stature or recumbent length, and maturity of the child and the statures of both parents (Roche *et al.*, 1975a[272]; Roche, 1980b[191]). It was clear also that the method should be sex specific. To be useful in a wide range of situations, the prediction method had to be applicable at many ages and when data from only one examination were available. The maturity measure had to be one that could be obtained at all childhood ages. Skeletal age is the only measure of maturity that meets this criterion.

Roche *et al.* (1975a[272]) used data from Fels participants who were at least 18 years old to develop their RWT stature prediction method. Data from examinations when skeletal maturation was complete or nearly complete were excluded because skeletal ages in years could not be assigned, and predictions made so late in relation to maturity would be of little practical value. Skeletal ages obtained by the atlas method were available for the hand–wrist, foot–ankle and knee, for individual bones and for combinations of these bones. These skeletal ages were a majority of the 80 independent (predictor) variables that were tested for inclusion in the final prediction equations. This set of possible predictor variables had to be considerably reduced. In the first steps principal components analysis and cluster analysis were used to exclude variables that could be represented by a linear combination of other variables. At the end of this process, it appeared that the same set of predictors could be used for all ages and both sexes. Fourteen variables were retained for further testing.

Recumbent length was chosen instead of stature because it can be measured at all ages. When only stature was available, it was adjusted to be approximately equivalent to recumbent length by adding 1.25 cm (Roche & Davila, 1974a[223]). The final selection of predictors was based, in part, on the view that variables should be easy to obtain and that radiation load and financial burdens should be minimized.

All possible subsets regression was used at each age: perhaps the first time this procedure was applied in a biological context. Area skeletal ages were shown to be more useful than bone-specific skeletal ages. Means and medians of groups of skeletal ages were about equally useful; medians were chosen because they are more resistant to effects of outlying values or adult levels of maturity in a few bones. The median skeletal age of the knee was slightly more useful than that of the hand–wrist, but hand–wrist skeletal age was chosen to reduce irradiation and to conform with common

Table 4.3. *Values for $R^2$ and selected percentiles for absolute differences (cm) between predicted and actual adult statures (data from Roche et al., 1975a[272])*

| Age (years) | Boys Percentiles | | | Girls Percentiles | | |
|---|---|---|---|---|---|---|
| | $R^2$ | 50th | 90th | $R^2$ | 50th | 90th |
| 1 | 0.58 | 2.74 | 5.94 | 0.49 | 3.30 | 5.85 |
| 2 | 0.59 | 2.50 | 6.25 | 0.44 | 3.00 | 5.80 |
| 3 | 0.67 | 2.26 | 5.91 | 0.52 | 2.80 | 5.46 |
| 4 | 0.67 | 2.16 | 5.64 | 0.56 | 2.55 | 5.17 |
| 5 | 0.73 | 2.05 | 5.65 | 0.56 | 2.27 | 5.72 |
| 6 | 0.74 | 2.09 | 5.75 | 0.60 | 2.20 | 5.53 |
| 7 | 0.71 | 2.06 | 5.82 | 0.62 | 1.91 | 5.55 |
| 8 | 0.75 | 1.83 | 5.52 | 0.69 | 1.88 | 4.47 |
| 9 | 0.76 | 1.80 | 5.61 | 0.62 | 2.30 | 5.07 |
| 10 | 0.77 | 1.59 | 5.05 | 0.57 | 2.13 | 5.46 |
| 11 | 0.73 | 2.03 | 5.66 | 0.62 | 2.18 | 5.42 |
| 12 | 0.70 | 2.29 | 5.24 | 0.59 | 2.32 | 5.64 |
| 13 | 0.61 | 2.58 | 6.55 | 0.70 | 2.01 | 4.45 |
| 14 | 0.63 | 2.92 | 6.35 | 0.75 | 1.87 | 4.32 |
| 15 | 0.70 | 2.15 | 4.78 | — | — | — |
| 16 | 0.91 | 0.97 | 3.20 | — | — | — |

practice. Consequently, the final predictors were the present recumbent length, weight, and hand–wrist skeletal age of the child, and mid-parent stature, which is the average of the statures of the two parents. Some of these predictors were unnecessary at a few ages, but the complete set was retained for all ages to ensure consistency.

In the boys, the $R^2$ values from the age-specific regression equations showed that the predictions were only moderately accurate at ages younger than 2 years, but the average of the $R^2$ values was high (0.8) from 0.5 to 15.5 years. The accuracy of the predictions was assessed, in addition, from the differences between the actual and the predicted adult statures (residuals). The median residuals for the boys were about 2.0 cm to 5 years, slightly less from 6 through 11 years and markedly less at ages older than 15 years (Table 4.3). The $R^2$ values were slightly lower in the girls. The median residuals for the girls were slightly greater than 3.0 cm to 2 years, between 2.0 and 3.0 cm to 6 years, and close to 2.0 cm from 6 to 10 years. Later the residuals decreased rapidly. The 95th percentiles of these absolute differences were about 5.0 cm at most ages in each sex.

Skeletal age was a much more important predictor for the girls than for the boys, particularly during pubescence, although the amounts of growth during pubescence are the same in each sex (Bock *et al.*, 1973). The sex difference in the importance of skeletal age as a predictor may be related to the greater growth after pubescence in boys than in girls (Roche, 1989a[207]). This may also be responsible for the greater decrease in the accuracy of predictions during pubescence for girls than for boys.

The weightings (coefficients) in the regression equations were smoothed across age. This allowed interpolations to ages between half-birthdays. This smoothing slightly reduced the accuracy of prediction in the Fels group but it was expected that it would increase the accuracy when the method was applied to other groups. It is difficult to smooth interrelated variables because the errors associated with smoothing are correlated unless precautions are taken.

To overcome this problem, the regression coefficients were transformed to unrelated (orthogonal) variables. After these transformed variables were smoothed, they were changed back to their original relationships. The smoothing was done with repeated running medians after which a polynomial was fitted. Figure 4.16 shows the progressive changes in the weightings for recumbent length in the girls from the unsmoothed state (a), to when they were smoothed by repeated running medians (b), to smoothing by a polynomial before (c), and (d) after they had been changed back to the original relationships.

These smoothed coefficients have biological interest in regard to the determinants of adult stature, but interpretation is difficult because of their interrelationships. The recumbent length weightings were positive and similar in the two sexes. They decreased from 3 to about 14 years, but increased at older ages. The weightings for body weight were negative at most ages, which is concordant with reports that obese children tend to become short adults. This is in agreement also with the finding that rapid increases in weight in some Fels participants were associated with reductions in their predicted adult statures (Roche & French, 1969[233]). The coefficients for weight were larger in the girls than in the boys, but, in each sex, they increased (larger negative values) to about 4 years and then decreased.

The coefficients for mid-parent stature were positive at all ages in each sex but were larger for the boys, which would be expected because of their greater adult statures. The coefficients for mid-parent stature decreased with age to about 8 years, after which there were marked increases in the boys but only slight increases in the girls. The skeletal age coefficients were near zero until 4 years in each sex. At older ages, they were generally negative and were larger in the girls than in the boys.

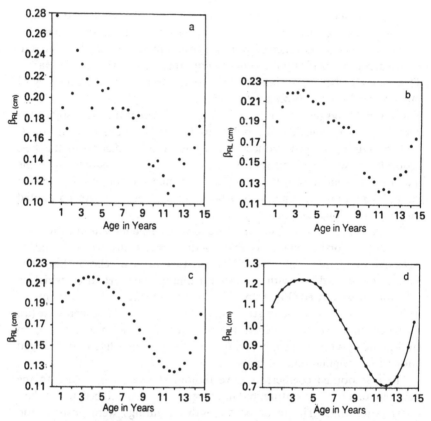

Fig. 4.16 Steps in the process by which the weightings (coefficients) for recumbent length in girls were smoothed during the elaboration of equations to predict adult stature. (Roche, A.F., Wainer, H. & Thissen, D. (1975a). Predicting adult stature for individuals. *Monographs in Pediatrics*, **3**, 1–115[272]. Redrawn with permission from S. Karger AG, Basel.)

The final prediction equations were presented for 1-month age intervals from 1 through 16 years (Roche *et al.*, 1975a[272]). The guidelines for their application stated that a population mean can be used when the value for a predictor variable is missing. If stature is measured instead of recumbent length, the recorded value can be adjusted to approximate recumbent length (Roche & Davila, 1974a[223]). Commonly, the measured stature of the father will be unavailable. A reported stature can be used or even the mother's categorization of his stature to short, medium, or tall (Himes & Roche, 1982). Even if the father's stature is completely unknown, a population mean can be substituted with only small increases in the errors of prediction (Wainer, Roche & Bell, 1978).

With a few exceptions, other variables did not add to the accuracy of the predictions. The growth patterns of the children, as described by the double logistic model of Bock *et al.* (1973), did not assist except that the age at the maximum rate of growth in the adolescent component was significantly correlated with the residuals (errors of prediction) in each sex. In addition, the predicted statures tended to be lower than the observed statures in late-maturing children.

Differences in the timing of the adolescent spurt, that were not accounted for by hand–wrist skeletal ages, were responsible for about 16% of the errors. This finding cannot be applied in prediction because the children would be too old for useful predictions when this age becomes known. It was postulated, however, that better estimates of hand–wrist skeletal age than the Greulich–Pyle values could help. Subsequently, the FELS method for the assessment of the hand–wrist (Roche, Chumlea & Thissen, 1988a[221]) was applied to children aged 15 years by substituting FELS skeletal ages for Greulich–Pyle skeletal ages when applying the regression equations. The mean errors were reduced by about 0.5 cm in each sex. This indicated that, at least during pubescence, the RWT prediction equations could be improved if new analyses were made using FELS skeletal age as a predictor.

The errors of prediction with the RWT method were compared with those from the Bayley–Pinneau method both for the Fels participants and participants in the Denver Growth Study. These comparisons were restricted to ages older than 6 years because the Bayley–Pinneau method is not applicable at younger ages. The median errors were much smaller with the RWT method than with the Bayley–Pinneau method in each group (Fig. 4.17). The only exceptions occurred near 16 years for boys in the Denver group. The RWT method can be applied only when a median Greulich–Pyle skeletal age for the hand–wrist is available. In the Fels group, these skeletal ages were calculated only when less than half the bones of the hand–wrist were adult. The data from the Denver cross-validation sample included all boys up to 16 years of age. Consequently, the RWT method was applied to some older boys in the Denver study for whom it was not appropriate.

Roche and Chumlea (1980[215]) analyzed serial RWT predictions of adult stature. The mean differences between predictions made 1, 2 or 5 years apart were near zero, but the standard deviations of these differences were about 1.2 cm, 1.5 cm and 2.0 cm respectively in each sex (Fig. 4.18). These standard deviations were smaller for intervals beginning at 3 to 9 years than for earlier or later intervals. These data showed the age range during which significant differences in predicted adult statures are most likely to

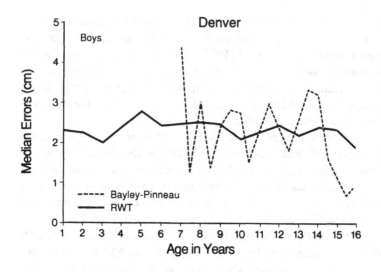

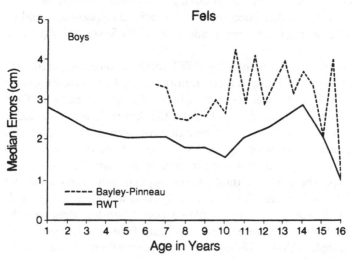

Fig. 4.17 Median differences between predicted and observed adult statures for boys in the Fels and Denver studies. (Roche, A.F., Wainer, H. & Thissen, D. (1975a). Predicting adult stature for individuals. *Monographs in Pediatrics*, **3**, 1–115[272]. Redrawn with permission from S. Karger AG, Basel.)

be detected in an intervention study. Additionally, these data can be used to analyze the significance of changes in predicted adult statures in association with therapy.

The prediction of adult stature has been reviewed in several publications in which attention was given to comparisons between methods (Roche,

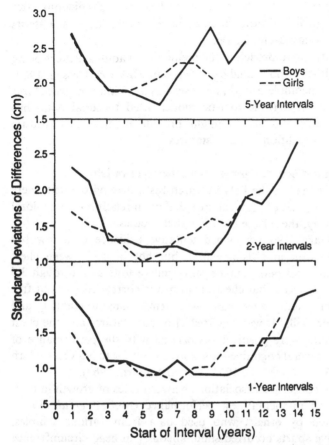

Fig. 4.18 Standard deviations of the differences between serial predictions of adult stature made 1, 2, and 5 years apart using the RWT method. (Roche, A.F. & Chumlea, W.C. (1980). Serial changes in predicted adult statures for individuals. *Human Biology*, **52**, 507–13[215]. Redrawn with permission from *Human Biology*.)

1980b[191], 1984b[197]). These comparisons were limited by the age ranges during which the methods are applicable. For example, only the RWT method can be used before the age of 4 years. The general conclusion from these comparisons, and those made by others, was that the RWT method is superior in regard to tendencies to under-predict or over-predict.

Random errors have been compared also. These were smaller and more consistent across age for the RWT than for the other methods. The only exception concerned older girls for whom the method of Tanner and colleagues has smaller random errors, presumably because it included separate equations for pre- and post-menarcheal girls. Predictions after menarche are, however, of little practical value, but this approach assists predictions for pre-menarcheal girls.

An entirely different approach to the prediction of adult stature is being investigated by Bock, who is analyzing Fels data (Bock & Thissen, 1980; Bock, 1982). This incomplete work, that uses Bayes' estimation procedures in combination with a triple logistic model fitted to serial data, has tremendous potential because it might provide predictions of future growth patterns in addition to adult statures.

### Associations between growth and behavioral variables

One might expect that, in the Fels Research Institute environment, there would be many analyses of physical growth data in relation to behavioral variables. Strangely, there have been few such studies.

Kagan and Garn (1963) provided tentative evidence of a low but significant correlation in girls (r = 0.2), but not boys, between chest breadth at 2 years and performance on cognitive tests at 2 through 3.5 years. The authors noted that chest breadth was correlated with fat-free mass and that this might be associated with items related to language and perceptual/motor skills. It was reported also that statures and the chest breadths of parents were positively associated with the performance of their children on several cognitive tests at ages from 2.0 to 3.5 years (Garn et al., 1960a[74]; Kagan & Moss, 1963; Kagan & Garn, 1963).

Another study concerned associations between rates of growth in head circumference and in mental test performance (McCall et al., 1983). Analyses reported by others, who used data from various samples, indicated that five spurts occurred simultaneously in head circumference and mental test performance between birth and 17 years. Furthermore, these earlier workers claimed that children were receptive to learning only during these spurts. Analyses of Fels data showed only one spurt for head circumference and three spurts in mental test performance between birth and 17 years. There were large differences between individuals in their patterns of change and between the patterns of growth in head circumference and those in mental test performance. It was concluded that there was no justification for changing school curricula because of supposed concurrent spurts in head circumference and mental test performance.

The literature concerning the growth of the brain, including that from the Fels study, has been reviewed critically (Roche, 1980c[192], 1981a[193]). The central topic addressed was the possible occurrence of 'catch-up growth' of the brain. The term catch-up growth refers to a period of rapid growth in an individual who was growing slowly. During the period of catch-up, the individual grows more rapidly than usual and, if catch-up is complete, the individual's growth status becomes normal. Catch-up growth is well documented for many body dimensions and organs but uncertain for the brain. Since brain size is related to cognitive development, catch-up growth of the brain would be very important in children treated for malnutrition and social deficits.

Increase in brain size is dependent on increases in the number of brain cells and in the size of these cells. It had been suggested that reductions in cell number are permanent. This is true for reductions early in gestation due to irradiation, virus infections, or chromosomal abnormalities. It may not be true for reductions at older ages due mainly to malnutrition. In malnourished fetuses, there are proportional decreases in the number and size of brain cells, but infants with small heads *in utero*, measured by ultrasound, tend to catch up incompletely in head size during the first year after birth. In older malnourished children, the decrease in cell size is greater than that in cell number. The deficits in cell size and in synaptic and dendritic complexity can be reversed.

It is difficult to determine the ages during which catch-up growth of the brain can occur. Some important studies of infants with good socio-economic status, who had intrauterine growth retardation or diseases that cause malnutrition that can be treated, e.g., pyloric stenosis, show incomplete catch-up in head circumference.

The reduction in head circumference in malnourished children is not entirely due to decreases in brain size and alterations in brain cells. As noted by Robinow (1968), there are decreases in scalp thicknesses, and in cranial bone thickness, but increases in the amount of the subarachnoid fluid that surrounds the brain. These changes are reversible and they may obscure alterations in brain size, judged from head circumference, during the rehabilitation of malnourished children.

There can be extraordinarily rapid catch-up growth in head circumference when some malnourished children are treated. This is associated with increases in intracranial pressure that cause sutural separation if they occur before the sutures are interlocked at about the age of 5 years. These increases probably involve brain size but it is not known whether the changes are in cells, myelin, or fluid. Rapid increases in head circumference are not possible after 5 years, but dendritic complexity, and perhaps other aspects of brain development, can increase even in old age. No longer can one accept the view that nutritional intervention of malnourished children is ineffective in regard to the central nervous system after the first 2 years of life.

One unusual study used noise exposure as an indicator of behavior and related this to growth and maturation. Siervogel and his associates (1982a) found that 24-hour noise exposure, measured with dosimeters, was not correlated with body size and maturity in boys after age effects were removed. In girls, however, weight, weight/stature$^2$, and measures of maturity were correlated with noise exposure. Larger and more mature girls, matched for chronological age, tended to be exposed to more noise.

Finally, Goodson and Jamison (1987) used age at peak height velocity and the percent of adult stature achieved at this age as measures of maturity. These measures were used to show that rapidly maturing adolescents were more concerned with their peers, had higher expectations for their futures, and were less independent and less interested in intellectual activities. When adult, individuals who matured rapidly were more concerned with power and proving their competence.

### Future directions

It is difficult to predict the long-term future of the Fels Longitudinal Study in regard to physical growth. As the Swedish proverb has it: 'The afternoon discovers what the morning never imagined.' Some short-term suggestions are possible, however, in keeping with the dictum of Winston Churchill: 'It is wise to look ahead but foolish to look further than you can see.'

Some possible future directions were included in a review by a National Institutes of Health panel chaired by Roche (1986a[201]). The following are some of the topics to which attention should be directed:

(i)     Reference data, including increments, are needed for the prenatal period. Incremental reference data would allow the early recognition of problems in prenatal growth that may indicate outcomes during the perinatal period and infancy. The extent and the timing of catch-up growth during infancy for fetuses with intrauterine growth retardation, across the range of severity and types of intrauterine growth retardation, are poorly known. Studies of change in fetal adipose tissue late in gestation are now possible due to improvements in ultrasonography.

(ii)    Reference data are needed for growth relative to maturity status during pubescence.

(iii)   There is a lack of studies of the associations of physical growth with mental and behavioral development although there are relevant data in longitudinal study archives. A fruitful field could be the analysis of changes in mental test performance in relation to hormonal changes during pubescence.

(iv)    Methods are required for the prediction of adult stature in abnormal children but it is difficult to obtain sufficient serial data from homogeneous groups of such children.

While these are worthy goals, their achievement would involve major extensions to the Fels study.

# 5  *Physical maturation and development*

Nature is comparatively careless of stature, permitting it to vary within relatively wide limits, but zealously keeps the program of maturation as nearly as possible to schedule.

*T. Wingate Todd* (1885–1938)

Maturation is the process that leads to the achievement of adult maturity. Maturation is a part of development. Both relate to progressive increases in complexity, but maturation is restricted to those developmental changes that lead to the same end point in all individuals. For example, the percentage of adult stature achieved by a child is a measure of maturity: all reach 100% in adulthood. Levels of maturity that are intermediate between the absence of measurable indicators and the adult state indicate the extent to which a child, or a group of children, has proceeded toward the completion of maturation in a particular body system. Maturation occurs in all body systems, organs, and tissues. For example, 'skeletal maturation' refers to a set of radiographically visible changes that culminate in the achievement of adult skeletal status by the early 20s.

This broad subject has received considerable sustained attention from Fels scientists. For example, Roche (1974c[180]) published an introduction to a symposium on adolescent physiology in which variation in maturation was the dominant theme. Much of this variation is associated with differences in rates of maturation among individuals. Attention was also given to the hormonal control of adolescence and the factors that regulate the timing of menarche. Maturation has also been the subject of a review by Chumlea (in press[13]).

This chapter will deal with: (i) skeletal maturation, (ii) dental maturation, (iii) sexual maturation, (iv) other aspects of maturation, (v) the non-invasive assessment of maturity, (vi) the regulation of maturation and its associations, (vii) the timing and sequence of adolescent events, (viii) the development of hearing ability, and (ix) developmental changes in skin reflectance.

### Skeletal maturation

Assessments of skeletal maturity (skeletal ages) are associated with body size and shape, the percentage of adult stature achieved, body composition measures such as percent body fat, bone widths, the widths of the cortex of bones, the timing of pubescence, and the age at which adult stature is reached (Garn *et al.*, 1961b[144]; Chumlea *et al.*, 1981a[24], 1983[39]). These values can identify children who are maturing rapidly or slowly and thereby indicate a need for laboratory investigations. Furthermore, these assessments are used in studies that describe and compare groups of children, and studies that evaluate the effects of environmental deficits. Additionally, skeletal maturity assessments are applied to predict adult stature and craniofacial growth (Roche, Chumlea & Thissen 1988a[221]). In summary, these assessments help describe and categorize children for clinical and research purposes. The general topic of skeletal maturation assessment was the subject of a symposium organized by Roche (1971c[176]).

The histological and radiographic changes during skeletal maturation have been reviewed by Roche (1986b[202]). Sequential alterations occur as the embryonic precursors of a bone gradually change until adult levels of form are achieved. A few of these alterations are shown diagramatically in Fig. 5.1. In A, a centrally placed marrow cavity is present with cancellous bone at the ends of this cavity and cartilage at each end of the model. B shows a later stage when the cells within the cartilagenous ends of the model enlarge prior to the formation of a center of ossification (epiphysis) in the cartilage. Stage C shows a well-formed epiphysis that is separated from the shaft of the bone by a layer of cartilage called the epiphyseal zone, beyond which there is a thin layer of 'compact bone' called the terminal plate. The adult stage is shown in D, where the epiphyseal zone is replaced by bone and there is bony fusion between the epiphysis and the shaft. Terminology is important. Places where ossification begins should be called 'centers of ossification,' not 'growth centers,' because the latter term implies they control subsequent growth, which is not the case.

Only some of the changes visible in radiographs meet the criteria for useful indicators of maturity. Some of the changes in epiphyses are important indicators, particularly those that involve alterations in epiphyseal shape and increases in the ratios between the widths of the epiphyses and the widths of the ends of the shafts. There are changes also in the bones of the wrist (carpals) which, after they ossify, gradually occupy larger

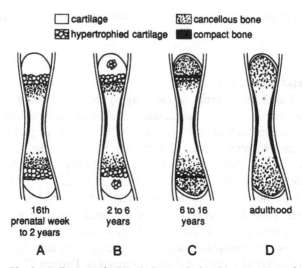

Fig. 5.1 A diagram of selected changes during the maturation of a long or short bone. Note the changes in the cartilage near the end of the bone that precede epiphyseal ossification (B), the formation of an almost fully ossified epiphysis (C), and the adult state (D).

proportions of the hand–wrist area as they change to become adult in shape. Epiphyses do not develop in the carpals but the changes in these bones are histologically and radiographically similar to those in the long and short bones. The long bones are those of the thigh, calf, upper arm, and forearm, whereas the short bones (metacarpals, metatarsals, phalanges) are in the palm, foot, fingers, and toes.

### Onset of ossification

The recognition of the onset of ossification in the epiphyses of long and short bones and in the carpals was the first method developed for assessing skeletal maturity. Onset of ossification has advantages as an index of maturity because the recorded data are not influenced by even wide variations in radiographic positioning, little observer training is required, and the data are highly reliable.

Counting the number of ossified centers is the simplest method of grading skeletal maturity, but this method provided insufficient information when it was applied to the carpal bones or the hand–wrist (Garn, 1960a[54]). Consequently, Sontag, Snell & Anderson (1939[302]) applied this method to the left side of the whole skeleton and reported the number of ossification centers present between birth and 5 years. Their distribution statistics for (i) all bones, (ii) metacarpals, metatarsals and proximal

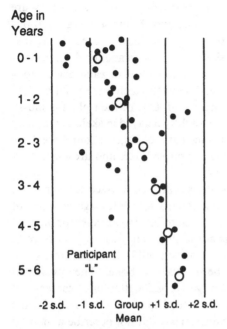

Fig. 5.2 Ages at onset of ossification of 38 centers plotted as z-scores (standard deviation levels). The open circles indicate means at annual intervals from 1 through 6 years for one Fels participant (data from Sontag & Lipford, 1943[288]). Note the tendency to more rapid maturation with advancing age in this participant.

phalanges, and (iii) the remainder showed a consistent advance of girls over boys that became marked at about 2 years of age.

Basing their views on earlier literature and on their own observations, Pyle and Sontag (1943) concluded that the timing of ossification in the hand–wrist may not be typical of that in the remainder of the skeleton. This topic received further attention at Fels during a long period, as will become evident later in this chapter. A further limitation was noted by Roche, Eyman and Davila (1971[230]) who showed that ages at onset of ossification were positively correlated with the later rates of skeletal maturation, which implied a negative correlation between the rates of maturation before and after the onset of ossification. Nevertheless, the number of centers ossified and atlas skeletal ages are highly correlated during infancy. This reflected the large contribution of onset of ossification to estimates of skeletal maturity with an atlas method in infancy (Pyle & Menino, 1939).

Sontag and Lipford (1943[288]) noted that variations in the sequence of onset of ossification were common and that some had considered these variations were mainly due to illnesses that delay centers when they are about to ossify, while not affecting other centers. Using serial radiographs of the major joints on one side of the body, they interpolated the ages at which 38 centers ossified and published means and standard deviations of these ages. They converted the age at appearance of each center to a standard score and plotted the standard score for each center that ossified during selected annual intervals in each child. One of their plots is shown in Fig. 5.2 for participant 'L' who was slightly delayed in skeletal maturity at 1 year, but progressively more advanced at older ages. From these data, Sontag and Lipford identified the most variable bones and drew attention to individual differences in variability.

Pyle and Sontag (1943) also showed that children who tended to mature rapidly had little dispersion of their standard scores for age at onset of ossification. These authors stressed the need to interpolate ages for onset of ossification when serial radiographs are studied. They wrote: 'We have ... estimated from serial roentgenograms on each child, the *actual* time when each cartilaginous center began to be replaced by bone.' They presented means and standard deviations of the ages at ossification of 61 centers and noted that ages at onset of ossification tended to be more variable for the carpals and tarsals than for other bones. They did not describe a method of skeletal maturity assessment, but their basic data were included in the formulation of the Fels Composite Chart (Sontag & Reynolds, 1945[298]), that was described in Chapter 4. This work by Sontag and his associates was biologically important, but because half the body had to be radiographed for its application, it did not lead to the development of a clinically applicable method.

In an interesting study, Robinow (1942b) applied factor analysis to determine whether there were subgroups among 19 bones in regard to their associations for the timing of onset of ossification. These bones 'represented' the arm and leg, centers that ossify early and late, and irregular and long bones. Using data from 31 children, Robinow identified two groups of bones with high correlations within each group, but low correlations with data for other bones in regard to age at onset of ossification. These groups were: (i) carpals and tarsals, and (ii) epiphyses and the patella. The head of the femur was not closely related to either group, and the trapezium was related to both groups. A general factor explained 24% of the total variance in age at onset of ossification while 36% of the variance was explained by the round bone (carpals, tarsals) and epiphyseal factors. Robinow noted that centers in the arm, leg, and those that ossify early or late did not form groups. Robinow interpreted these findings as justification for assigning one skeletal age to a whole hand–wrist, but he

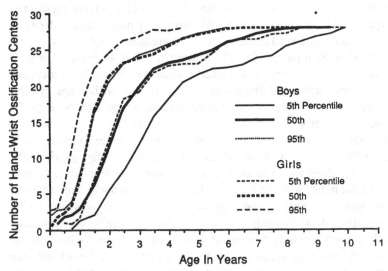

Fig. 5.3 Selected percentiles for the number of centers ossified in the hand–wrist at various ages in boys and girls (data from Garn & Rohmann, 1960b[135]).

warned that this age would be misleading in children with large differences in maturity levels between the carpals and the other bones of the hand–wrist. He recommended that, at least in these children, separate skeletal ages should be assigned to the carpal bones and to the other hand–wrist bones. An interesting aspect of this work is its sophisticated statistical nature, given that it was reported in 1942.

The lack of data relating to maturation during infancy was earnestly lamented by Falkner (1971b). Many such basic data have been provided, however, by Sontag and others (1939[302]), as noted earlier, and by Reynolds and Asakawa (1951) who presented tables of the number of centers ossified in the hand and foot from birth to 3 years. The Sontag group reported the distributions that Falkner sought, but Reynolds & Asakawa provided only ranges for children grouped according to five arbitrary rates of maturation. Reynolds and Asakawa (1951) provided data for ages at onset of ossification of selected centers near the hip, elbow, knee and shoulder, and they reported that the hand–wrist differed from the mean of the other areas by two categories in 1% of cases and by one category in 29% of cases. The foot–ankle was only slightly less representative of the total body than the hand–wrist. Their rapid method for the clinical assessment of the maturity level of the hand and foot was based on too few indicators to be precise.

Using a similar approach, Garn and Rohmann (1960b[135]) reported distributions of the number of hand–wrist centers ossified at particular ages from 1 month to 10 years. The 5th, 50th, and 95th percentiles showed the expected advancement of girls over boys at 1 month and older (Fig. 5.3). Percentiles were used because such data are commonly skewed (Garn, Blumenthal & Rohmann, 1965g[71]). During the first 4 years, there were rapid increases in the number of ossified centers in each sex. Later increases with age were slight and there was little variation within age groups, indicating that the number of ossified centers would not be a useful discriminator after 4 years.

Serial data showed little tendency for children to retain their percentile levels for the number of ossified centers during the first few years of life. Additionally, there were low correlations between the number of ossified centers in the hand–wrist and the number in the remainder of the arm and the leg, showing that, in this respect, the hand–wrist was not highly representative of the skeleton. From these findings, Garn and Rohmann (1960a[134]) concluded that the number of ossified centers in the hand–wrist, while not precise, was useful for group studies and clinical work between birth and 4 years. This view was based, in part, on their earlier finding that the ages at ossification for the hand–wrist centers that ossify early were only slightly correlated with the ages at ossification of those that ossify later (Garn & Rohmann, 1959[133]).

The onset of ossification in 28 hand–wrist centers was analyzed by Garn and Rohmann (1959[133]). Having transformed the data to remove skewness, they calculated correlations between bones in their ages at onset of ossification and used the means of these correlations as communality indices for each bone. These indices were interpreted as measures of the extent to which timing in one bone was representative, or predictive, of timing in the group. This study ushered in an important new concept of the need to weight bones differentially when assessing skeletal maturity.

Garn and Rohmann (1959[133]) reported that girls tended to have higher communality indices than boys which they suggested might be due to redundancy of X-chromosomal material in girls. In each sex, the carpals, radius and ulna had relatively low communality indices in each sex, as did the first proximal phalanx in girls. Garn & Rohmann concluded that the carpals had less predictive power than the other bones of the hand–wrist in regard to the onset of ossification and, therefore, they were of little value in the assessment of skeletal maturity. Contrariwise, the proximal and distal phalanges and the metacarpals had high communalities for ages at onset of ossification of hand–wrist bones, and, therefore, indicated the general level of maturity in the hand–wrist.

A corresponding study of the foot was reported by Garn and Rohmann (1966b[142]). The proximal phalanges and metatarsals had relatively high communality indices with other bones for ages at onset of ossification. As for the hand–wrist, communality indices in the foot tended to be higher for girls than boys; the values of these indices were similar for the hand and foot.

These analyses were extended when Garn *et al.* (1966a[146]) showed that communality indices were higher between joints within limbs than between either homologous (e.g., elbow, knee) or non-homologous (e.g., elbow, hip) joints in different limbs. These authors postulated that coordinated development within limbs may have favored evolutionary selection pressure. Garn and his colleagues also found a pattern of low communalities between bones that differed markedly in the timing of onset of ossification.

In another study, age at onset of ossification of the adductor sesamoid was shown to be highly variable, with median ages of about 13 years for boys and 11 years for girls (Garn & Rohmann, 1962b[137]). The adductor sesamoid is the name given to a bony nodule that forms in a tendon at the base of the thumb. The timing of adductor sesamoid ossification was not highly correlated with that of other bones that ossify at about the same age, but it was correlated with age at menarche and with age at epiphyseal fusion at the proximal end of the tibia ($r = 0.6$). Apparently, maturation of the adductor sesamoid is more closely allied to that of long bones than to maturation of the carpals. This is one of the few reports in which Garn and his colleagues considered external relationships of ages at onset of ossification. The need for such consideration was stated by Reed in a *Symposium on Assessment of Skeletal Maturation* (Roche, 1971b[175]). In discussing the assignment of numerical values to stages of skeletal maturation, he stated that these were commonly based on inter-relationships between bones and that stages with low internal relationships were excluded. Reed considered this approach was risky because the stages excluded may be closely related to important external characteristics such as illness.

Increments in the number of hand–wrist centers through 7.5 years were reported by Garn *et al.* (1961c[152]). Many of the distributions of these increments were markedly skewed and there was considerable variability to 3 years. Garn and his co-workers noted large differences between the trends for individuals and for the group which limited the practical value of these increments. These increments were not significantly correlated with increments in stature and weight.

A 'rational approach to the assessment of skeletal maturation' was described by Garn and his colleagues (Garn, Silverman & Rohmann, 1964e[167]; Garn *et al.*, 1966a[141]). They calculated an intercorrelation matrix of the ages at onset of ossification in 52 centers in the hand–wrist and foot–ankle. Basing their judgments on communality indices (means of logs of correlation coefficients), they selected the 19 centers that were most informative about the ages at onset of ossification of other centers in the hand and foot. Later, they modified this approach when they identified the bones that were most informative about the remainder of the skeleton. Their results were unlikely to have been influenced by familial atypical ossification sequences because bones subject to these were excluded by their selection process. For example, one wrist bone, the triquetral, was excluded because it may ossify unusually early in some families without any relationship to ossification timing as a whole. Other centers, such as the middle phalanx, were excluded because commonly their epiphyses do not ossify.

Garn *et al.* (1966a[141]) stated that it was difficult to develop a method of skeletal age assessment because skeletal maturity levels varied between parts of the skeleton, particularly soon after birth and near the end of growth. They concluded that the bones of the hand–wrist, elbow, knee and foot–ankle had similar communality indices, but the hand–wrist may be more useful than the other areas because it contains many bones. This led Garn and his colleagues to recommend that skeletal maturity be assessed from radiographs of the hand–wrist, foot–ankle and knee for an overall appraisal, but that a particular area be assessed when there is specific interest in its maturity. This view is based on the assumption that communality indices for onset of ossification match those for later stages of maturation.

Garn, Rohmann and Silverman (1967e[154]), summarizing much of their earlier work, presented reference data for the timing of postnatal ossification. In this publication, they reported later timing for boys than girls and that the sex differences in timing varied by bone whether expressed in years or as ratios (male age/female age). Consequently, separate sets of reference data are needed for each sex and accurate estimates of skeletal maturity must be derived from bone-specific or indicator-specific data. They noted: 'Only now are we achieving an understanding of how to use the information from radiographs for the best possible appraisal of skeletal development.'

Garn and his colleagues (1967e[154]) selected the 20 centers with the highest communality indices for ages at onset of ossification in the major joint areas. This selection led them to modify their earlier recommendation; they suggested that both the hand–wrist and foot–ankle be radiographed to assess skeletal maturity. They claimed that their 20 centers provided a rational basis for the development of a 'point additive' system of skeletal age assessment. They did not extend this approach to the development of a complete system for the stages of skeletal maturation that follow the onset of ossification.

The prevalence of missing ossification centers for the bones of the foot was documented by Garn, Rohmann and Silverman (1965e[153]). This is relevant to the assessment of maturity because an assessment method must be based on features that occur during the maturation of all normal individuals. Absence (agenesis) was more common in girls than boys. Siblings were alike in the number of missing centers, with closer concordance for sisters than brothers, which was consistent with X-linked dominant inheritance. In the foot, a center for the fifth middle phalanx was missing in about 98% of children and a center for the fourth middle phalanx was missing in 54% of boys and 70% of girls. The tendency to agenesis was a progressive phenomenon in which centers were eliminated medially and distally from an epicenter in the fifth toe. When a phalangeal center was missing, that toe tended to be short, with a narrowing of joint spaces or fusion between its segments.

### Sequence of onset of ossification

The sequence of onset of ossification may be associated with rates of growth and maturation, and with illness experience, but variability in these sequences is masked when only mean ages at onset are reported. Serial data are needed to establish the sequences for individuals. Pyle and Sontag (1943) claimed that the sequence of onset of ossification was unusual for the carpals and tarsals when the rate of growth was either rapid or slow. This possible association has not received further attention. Unusual sequences of ossification in the hemiskeleton were said to be more common in slowly maturing children because they are more likely to be recognized in such children who have more radiographs than other children during a particular phase of maturation. Additionally, they are more likely to be observed when serial radiographs are closely spaced and when there are small differences between a pair of centers in their median ages of ossification. Nevertheless, unusual sequences of hand–wrist bones were not associated with the rate of skeletal maturation (Garn et al., 1966a[146]).

Unusual sequences were not associated with more frequent illnesses in girls, but there was a slight tendency to such an association in boys, even after the data for one boy with many illnesses and an unusual sequence were omitted (Garn et al., 1966a[146]). In a review article, Garn (1966c[68]) postulated that malnutrition may alter the order of appearance of ossification centers, but the prevalence of sequence variability in the hand–wrist was similar for moderately malnourished Guatemalan children

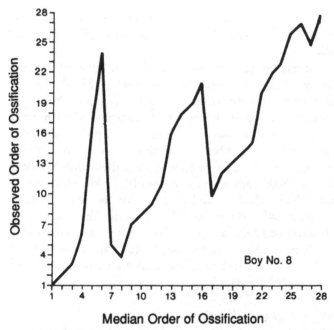

Fig. 5.4 An aberrant sequence of ossification in hand–wrist centers. The numbers on the horizontal axis indicate the median sequence; those on the vertical axis show the observed sequence in one Fels boy. (Garn, S.M. & Rohmann, C.G. (1960b). Variability in the order of ossification of the bony centers of the hand and wrist. *American Journal of Physical Anthropology*, **18**, 219–28[135]. Redrawn with permission from Wiley-Liss, Inc.)

and for Fels participants (Yarbrough *et al.*, 1973). The data of Garn *et al.* (1966b[142]) suggest, however, that malnutrition may cause small changes in the median sequence for boys, but there was no evidence of a similar effect for girls.

Garn and Rohmann (1960b[135]), using data from serial hand–wrist radiographs, analyzed variability in the sequence of ossification within individuals. An example of their data is shown in Fig. 5.4 in which the horizontal axis shows the median order of ossification for the group and the vertical axis shows the observed order for one boy. The centers that are ordered 8 and 17 in the group sequence ossified unusually early and the center that is ordered 6 for the group ossified unusually late. In this boy, an analysis of alternate sequences of ossification, such as: 'A' before 'B,' 'B' before 'A,' for 18 hand–wrist centers with similar median ages of ossification showed the number of associations between bones for alternate sequences greatly exceeded chance, particularly for metacarpal and phalangeal centers, although a single determinant did not explain all of them (Garn, Rohmann & Wallace, 1961d[158]). These workers also noted that unusual sequences of onset of ossification tended to be familial (Garn & Shamir, 1958[164]; Garn & Rohmann, 1960b[165]; Garn et al., 1961c[152]).

The sequence of onset of ossification in the wrist was shown to be variable and it differed between the sexes (Garn et al., 1966b[145]). In girls, the triquetral tended to be later in the sequence, and it was less common for the trapezium to precede the trapezoid. Marked variations in sequence were common in the foot; the differences in sequence between the sexes were sufficient to require sex-specific reference data. Additionally, differences between the sexes in the sequence of ossification for the elbow, shoulder and knee necessitated the use of sex-specific reference data (Garn & Rohmann, 1960b[165]; Garn et al., 1966b[142], 1967e[154]).

The many studies of the sequence of onset of ossification and of associations between bones in ages at onset of ossification made by Garn and his colleagues could be expanded to maturity indicators that develop after the onset of ossification, but before epiphyseal fusion. This would be a huge task, and it may not serve a major need. The work done by Garn did not lead directly to a practical method for the assessment of skeletal maturity, but it provided much necessary basic information and important concepts. For example, Garn and his colleagues concluded that centers with marked common sequence variability should be omitted when skeletal maturity is assessed, although this diversity in sequence is part of the normal population variability (Garn et al., 1966b[145]).

Citing variability in timing and sequence, Garn and Rohmann (1960a[134]) expressed skepticism that a skeletal age assessment method could be based on the enumeration of centers. Yarbrough et al. (1973), however, described a method for pre-school children that was developed from Guatemalan data and validated using Fels data. It was shown by Yarbrough and his colleagues that the *number* of ossified centers in the hand–wrist allowed one to predict *which* centers were ossified with only small errors. Apparently, sequence variation is not of practical importance in the assessment of skeletal maturity. Furthermore, the number of centers ossified provided more accurate identification of the particular centers ossified than did the atlas method or the Tanner–Whitehouse method. The atlas and Tanner–Whitehouse methods utilize additional information but, at ages up to 6 years, they are strongly influenced by onset of ossification. This interesting approach was not developed further, mainly because it is not applicable at older ages.

### Epiphyseal fusion

Fels scientists have given considerable attention to epiphyseal fusion which is the final stage in the maturation of a bone that has an epiphysis (Stage D in Fig. 5.1). Epiphyseal fusion is not detected as easily as the onset of ossification. Bones that are elongating rapidly may be wrongly supposed to be in an early stage of fusion due to the superimposition of the diaphyseal and epiphyseal outlines (Garn *et al.*, 1967e[154]). Epiphyseal fusion is an important stage of maturation. It is part of growth dogma that elongation of a long or short bone does not occur after the epiphysis is fused to the shaft (diaphysis). Hertzog (1990), however, has presented evidence for slow elongation of epiphyseal bones in Fels adults; these changes presumably occurred at the non-epiphyseal ends of these bones (Roche, 1965[173]).

'Fields' of hand–wrist bones, in regard to the timing of epiphyseal fusion (Fig. 5.5), were described by Garn *et al.* (1961b[144]). These fields included groups of bones within which the timing of fusion was highly correlated. These fields were: (1) distal phalanges I–V; (2) middle phalanges II–IV and proximal phalanx I; (3) proximal phalanges I–V and metacarpal I, and (4) metacarpals II–V. They found that the timing of onset of ossification and of epiphyseal fusion were not associated, although the timing of both these events appeared to be partly gene related. Consequently, they concluded that skeletal maturation may not be a unitary phenomenon.

The timing of epiphyseal fusion in the hand–wrist is closely correlated ($r = 0.9$) with age at menarche and with age at fusion of the proximal epiphysis of the tibia (Garn *et al.*, 1961b[144]). Garn and his colleagues made another very important observation: although the median ages of occurrence were similar for fusion at the proximal end of the tibia and the completion of fusion in the hand–wrist, the sequence of these events varied with differences in timing that may be as long as 2 years in each direction (Fig. 5.6). This illustrated the limitations of skeletal age assessments that were made for the whole skeleton from a radiograph of one area. Garn and his co-workers suggested that these variations in timing may explain why growth in stature continued in most children, but not all, after epiphyseal fusion in the hand–wrist was complete.

### The atlas method of assessment

Several studies at Fels have utilized the atlas method of assessing skeletal maturity. This method is based on pictorial standards with which radiographs are compared. The best atlases were produced in Cleveland (Ohio) where many serial radiographs of normal children were available.

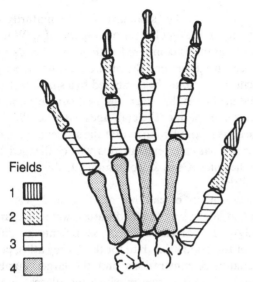

Fields

1

2

3

4

Fig. 5.5 Fields of hand–wrist bones in regard to the timing of epiphyseal fusion. (Data from Garn, Rohmann & Apfelbaum, 1961b[144].)

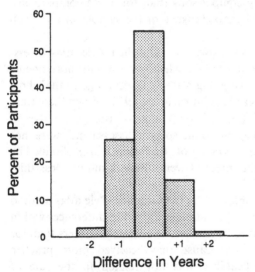

Fig. 5.6 The timing of the completion of fusion at the proximal end of the tibia relative to the completion of epiphyseal fusion in the hand; sexes combined. (Data from Garn, Rohmann & Apfelbaum, 1961b[144].)

Groups of radiographs for children the same age were ordered according to the maturity levels of one bone at a time. Subsequently, radiographs were selected that represented the central tendency of each bone and one of these radiographs was chosen as the 'standard' for the maturity of the area, e.g., hand–wrist, at the particular chronological age. When this method is applied, a radiograph is compared with the standards for one bone at a time, but those who developed the method did not specify how the resulting bone-specific ages should be combined to a single skeletal age for the hand–wrist. The median is, however, the best simple summary of the information from bone-specific skeletal ages (Roche, Wainer & Thissen, 1975d[274]). The atlas method implied that the sequence of maturity indicators was fixed in normal children, but this was challenged by the findings of Garn and colleagues (Garn & Rohmann, 1960b[135]; Garn et al., 1961b[144], 1966b[142], 1967e[154]).

Paired hand–wrist assessments of children aged up to 5 years, made using the Florey atlas and later using the Todd atlas, were compared by Pyle and Menino (1939). The correlations for assessments by different observers were higher for the Todd atlas than for the Florey atlas, perhaps reflecting the better quality of illustrations and the larger number of standards in the Todd atlas, and the order in which the atlases were used. The Fels children tended to lag behind the standards in both these atlases, with larger deficits for Florey comparisons than for Todd comparisons. These differences reflected the characteristics of the samples from which the atlases were derived.

In other investigations it was shown that Greulich–Pyle atlas assessments of the hand–wrist were more replicable if bone-specific assessments were made instead of directly assigning a single skeletal age to the whole hand–wrist (Roche et al., 1970a[228]; Johnson et al., 1973). When there were wide ranges of maturity within individual hand–wrists, direct overall assessments were less reliable, but bone-specific assessments were not affected. Furthermore, it is important to note that skeletal ages obtained by the recommended bone-specific method were about 2 months less than those made by the overall method.

Bone-specific skeletal ages obtained by the Greulich–Pyle atlas method were more replicable at younger than at older ages. The differences within and between observers tended to be smaller for research workers than for pediatric radiologists, probably because they received more practice (Johnson et al., 1973). Replicability was not related to the rate of maturation, and changed only slightly when the carpal bones were excluded (Roche et al., 1970a[228], 1970b[263]). The carpals varied markedly in rates of maturation, which could be an advantage if they provided valuable discriminant data or information about environmental influences. Training was a major factor in replicability. Differences between observers were comparatively large when training was limited to reading the atlas, but when it included assessing 30 radiographs, and discussing the assigned skeletal ages with an experienced person, reliability was similar to that for established assessors.

Later, Roche and Davila (1976[226]) analyzed the reliability of Greulich–Pyle atlas assessments in an unusual way. Repeated assessments were made of radiographs that were completely masked except for one bone. After this bone had been assessed in each radiograph, the masking was changed so that a different bone was visible. This approach provided independent skeletal ages for each bone that were compared with bone-specific assessments made when the whole of each radiograph was visible. Replicability was higher and the ranges of bone-specific skeletal ages within radiographs were much narrower for assessments made when the whole hand–wrist was visible, but replicability for bone-specific skeletal ages was lower, especially for the hamate and triquetral. It was concluded that when all the bones were visible the skeletal ages assigned to bones assessed late in the sequence were influenced by the ages assigned earlier.

Pairs of radiographs at different chronological ages were assessed by the Greulich–Pyle and Tanner–Whitehouse methods. Predictions of skeletal ages at older chronological ages, using skeletal ages at younger chronological ages as predictors, were more accurate with the Greulich–Pyle method than with the Tanner–Whitehouse method (Roche *et al.*, 1971[230]). The errors of prediction using linear trends fitted to serial data were greater than those from single measurements, unless many serial data points were available. These scientists described a simple method of prediction from a single value that was based on the assumption that the standard deviation score will be retained.

In another study, Himes (1977) reported that Greulich–Pyle and Tanner–Whitehouse skeletal ages were correlated more closely with ages at peak height velocity than with ages at menarche and that these correlations were consistently higher for Greulich–Pyle than for Tanner–Whitehouse skeletal ages.

### The Roche–Wainer–Thissen (RWT) method for the knee and the FELS method for the hand–wrist

An early account of the statistical basis that led to the development of the RWT method for the knee and the FELS method for the hand–wrist (Roche *et al.*, 1975d[274], 1988a[221]) was published by Murray, Bock and Roche (1971). This was the first paper to suggest the use of maximum likelihood procedures and item response theory for the assessment of skeletal age. This major methodological advance is likely to replace the atlas method. In the RWT and FELS methods, a logical mathematical process is used to combine the data from the various indicators to a single estimate of skeletal maturity for a child together with the standard error (confidence limit) of this estimate (Thissen, 1989).

The RWT method for the assessment of the knee was developed using almost 8000 radiographs (Roche *et al.*, 1975d[274]) and about 14000 radiographs were used to develop the FELS method for the hand–wrist. These radiographs were taken during several decades but there are no significant associations between the skeletal ages that were assigned and years of birth. Other analyses made by comparing skeletal ages of parents and their children of the same sex, when both were assessed at the same chronological age, showed only small irregular differences. Correspondingly, differences in age between mothers and daughters at menarche were near zero.

It is well known that boys mature more slowly than girls and, therefore, grades of brief duration are more likely to be observed in boys than in girls. This made it desirable to develop the new methods for boys and then apply them to girls. The work began with screening in which a few radiographs (primary test group) were used to determine which potential maturity indicators should be excluded because they were clearly not useful. The process included replicate grading by two independent assessors. Each radiograph was assessed without reference to earlier or later radiographs of the participant. An indicator was retained for further testing if it discriminated within chronological age groups, its most mature grade was universal at older ages, and it was reliable, valid, and complete.

Reliability was determined from differences in the grades assigned by different assessors and the differences when grading was repeated by the same assessor (inter- and intra-assessor differences). Validity, or the ability of the indicator to provide information about maturity, was judged from the frequency with which serial assigned grades changed illogically from more mature to less mature (reversals) with advancing chronological age. It was also determined whether each indicator had the quality of completeness, i.e., whether it could be assessed in all radiographs during the age range for which it was applicable.

After this primary testing, a group of serial radiographs of boys was used for further tests of the indicators that were retained. After this testing, the final indicators were chosen and graded on all the radiographs of the boys and the girls. The data from the assessments of all the radiographs were used to plot the changes in the prevalence of grades in relation to chronological age. Logistic functions fitted to these data, using maximum likelihood procedures, provided estimates of the age at which the prevalence of each indicator grade reached 50% (threshold), the rate of

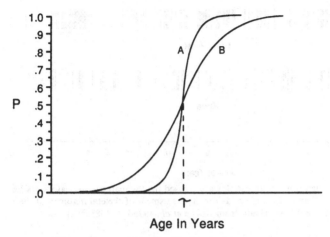

**Age In Years**

Fig. 5.7 Two theoretical indicators that have different slopes, but reach their thresholds (50% prevalence) at the same age. (Roche, A.F., Wainer, H. & Thissen, D. (1975d). *Skeletal Maturity: The Knee Joint as a Biological Indicator*. New York: Plenum Publishing Corporation[274]. Redrawn with permission from Plenum Publishing Corporation.)

change in prevalence with age (slope), and the regularity of the changes in prevalence.

Indicator grades may reach their thresholds at the same age but have different slopes (Fig. 5.7). The steeper the slope, the more rapid the change in prevalence with age, and the more informative the indicator. Some indicator grades had identical slopes but reached their thresholds at different ages. These indicators were equally informative, but they differed in the ages of greatest utility which are near the thresholds. Ideally, the ages of the thresholds would be evenly distributed across the age ranges during which the methods can be applied. This need was approximately met, but many of those for the hand–wrist occurred after 15 years and many of those for the knee occurred before 6 years (Fig. 5.8; Roche *et al.*, 1988a[221]).

The RWT and FELS methods were based on many maturity indicators, but at a specific chronological age only a few need be graded because each has a limited age range during which it is useful. The slopes and thresholds from each indicator are combined to a single index of maturity (estimated skeletal age) using techniques derived from item response theory, as developed for psychological measurement. This index was scaled in years with the means equivalent to the corresponding values for chronological age.

This Chapter began with a well-known extract from the writings of T. Wingate Todd who dominated this subject area during the 1920s and

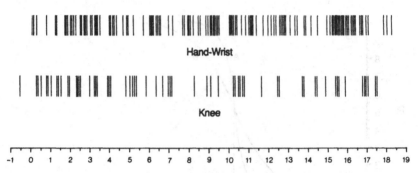

Fig. 5.8 Ages at which thresholds are crossed for the maturity indicators graded in the RWT and FELS methods for the assessment of skeletal maturity of the hand–wrist and knee. (Data from Roche *et al.*, 1975d[274], 1988a[221].)

1930s. This statement is expressed beautifully, but it is incorrect. These new methods for the assessment of skeletal maturity made it possible to adjust the observed variability in skeletal age for the errors of the estimates. Consequently, it could be shown that skeletal maturity was about twice as variable as stature.

The RWT and FELS methods provide the standard error of each estimate of skeletal age. The standard error of an individual estimate is small when many informative indicator grades are assessed with thresholds near the chronological age of the child, and it is large in the opposite circumstances. The errors were larger at some ages than others as shown by the total information curves for the knee and hand–wrist (Fig. 5.9). A relatively large amount of information was provided by a knee radiograph until 2 years, after which it decreased to 6 years and then remained approximately constant. The amount of information from the hand–wrist was similar to that from the knee until 3 years, but the hand–wrist was more informative than the knee after this age, despite decreases at ages after 14 years. The standard errors of the RWT and FELS methods were higher at ages older than 14 years because fewer maturity indicators can be assessed than at younger ages. There were only small sex differences in the amounts of information provided. The RWT and FELS methods allow for missing values, which is important if part of an area cannot be measured.

The RWT and FELS assessments were highly replicable (Table 5.1), with smaller assessor differences than those generally reported for other methods. There were, however, some large differences between the means for skeletal ages obtained with the atlas methods and those obtained with the RWT and FELS methods (Roche *et al.*, 1975d[274], 1988a[221]; Chumlea, Roche & Thissen, 1989a[37]). These reflected variations between the groups

Table 5.1. *Replicability of RWT and FELS assessments of skeletal ages in years (data from Roche et al., 1975d[274], 1988a[221])*

|  | Percentiles (years) | | |
|---|---|---|---|
|  | 10 | 50 | 90 |
| *RWT (knee)* | | | |
| Interobserver differences | 0.02 | 0.26 | 0.80 |
| Intraobserver differences (A) | 0.0 | 0.18 | 0.71 |
| Intraobserver differences (B) | 0.0 | 0.08 | 0.59 |
|  | Mean | SD | |
| *FELS (hand–wrist)* | | | |
| Interobserver differences | 0.20 | 0.21 | |
| Intraobersver differences (A) | 0.08 | 0.08 | |
| Intraobserver differences (B) | 0.17 | 0.13 | |

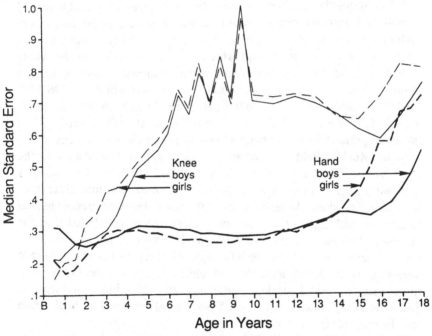

Fig. 5.9 The information available from the hand–wrist and knee at various ages as shown by the median standard errors. Small standard errors indicate relatively large amounts of information. (Roche, A.F., Chumlea, W.C. & Thissen, D. (1988a). *Assessing the Skeletal Maturity of the Hand–Wrist: FELS Method*, pp. viii + 339[221]. Redrawn, courtesy of Charles C Thomas, Springfield, Illinois.)

against which each method was scaled. Scaling for the RWT and FELS methods was appropriate for a method to be used in the US since the distributions of skeletal ages in the Fels Longitudinal Study are similar to those from US national surveys (Roche, Roberts & Hamill, 1974c[260], 1976a[261]).

Correlations between RWT skeletal ages and skeletal ages obtained by other methods ranged from 0.4 to 0.8, and the coefficients of reliability between FELS and other hand–wrist skeletal ages varied from 14% to 89%, which showed considerable independence (Roche *et al.*, 1975d[274], 1988a[221]).

### Choice of a skeletal area for the assessment of maturity

The part of the skeleton chosen for the assessment of skeletal maturity should be one that is easy to position for radiography, and minimal radiation should be required. As few radiographic views as possible should be obtained, and all useful relevant information should be extracted from each radiograph. In general, the hand–wrist is the best area, but there are limitations to the accuracy with which one can generalize to the whole skeleton from assessments of the hand–wrist, as shown by analyses of ages at onset of ossification and skeletal maturity levels (Garn *et al.*, 1961b[144], 1964e[167], 1966b[145], 1967e[154], 1967f[147]; Roche & French, 1970[234]; Roche *et al.*, 1975d[274], 1988a[221]). Much earlier, Sontag *et al.* (1939[302]) warned against the assumption that 'as the hand grows, so grows the entire skeleton.'

Xi and Roche (1990) reported an analysis of paired skeletal ages for the hand–wrist (FELS method) and the knee (RWT method). They showed that, within age- and sex-specific groups, the mean absolute differences between paired skeletal ages were about 0.5 years. These differences tended to increase with age until an age of 9 years. It was concluded that the observed differences were too large to be entirely due to errors of measurement and that the skeletal ages of these two areas were not interchangeable. In at least 5% of children, the choice of an area for assessment would markedly influence an evaluation. Furthermore, descriptions of populations that are based on only one part of the skeleton may be misleading.

There has been considerable uncertainty as to whether the carpals should be included when the skeletal maturity of the hand–wrist is assessed. These bones are variable in their rates and patterns of maturation (Garn & Rohmann, 1959[133], 1960b[135]). This variability may make it more difficult to construct methods for their assessment, but their inclusion would be an advantage if it assisted meaningful discrimination between individuals. Their inclusion reduced the standard errors and, therefore, it increased the stability of the estimates from 7 to 13 years in boys and 4 to 10 years in girls (Roche, 1989c[209]). Therefore, the carpals should be assessed during these age ranges.

Brother–sister correlations for FELS hand–wrist skeletal ages were generally lower than those for other types of sibling pairs, but they were similar to the brother–brother correlations after 11 years. After 12 years, the sister–sister correlations were considerably higher than the others, suggesting an X-chromosome influence at these ages (Xi, Roche & Guo, 1989b).

## Dental maturation

Dental maturation has received considerable attention from Fels scientists who have based their analyses on the emergence of the crowns of teeth into the mouth and on maturational changes in teeth that can be observed in radiographs. Very useful reference data for ages at emergence of the deciduous teeth were provided by Robinow, Richards and Anderson (1942). After combining data from the two sides, these workers plotted frequency distributions for each sex that showed slight skewness to the right. There were significant tendencies for the lower incisors to emerge before the upper incisors and for emergence to occur earlier and be less variable in timing in girls than boys. After converting the values to standard scores, these workers calculated an intercorrelation matrix of ages at emergence, the ages of the children when their statures were equal to the median stature at 18 months in a reference group and their ages when the number of centers ossified in the extremities were equal to the median number ossified at 18 months in a reference group. Factors extracted from the matrix showed ages at emergence were highly correlated within but not between groups of anterior and posterior teeth and that ages at emergence were not closely related to the other indices of maturity that were studied. Correspondingly, there were only low correlations between the ages at which most dental maturity stages were reached and either stature, weight, number of hand–wrist centers ossified or Greulich–Pyle atlas skeletal ages (Lewis & Garn, 1960). Nevertheless, the correlations between the timing of late stages in the calcification and the emergence of the second mandibular molar tooth and the timing of either menarche or epiphyseal fusion at the proximal end of the tibia were higher (0.3 to 0.6).

In studies of fossils, dental emergence is commonly used as a taxonomic criterion, but the data for emergence differ from those in modern man (Koski & Garn, 1957; Garn, Koski & Lewis, 1957b[97]; Koski, Garn & Lewis, 1957; Garn & Lewis, 1958[99]) in whom 'emergence' refers to the appearance of part of a tooth through the gum. Some have established fossil emergence sequences by extrapolation from the stages of maturity of unerupted teeth. Fels data show that the permanent teeth that succeeded the deciduous teeth emerged through the gum soon after the deciduous precursor was lost, but the permanent molar teeth did not emerge through the gum until they reached the occlusal level.

In significant papers, Garn *et al.*, (1956a[124]) and Garn, Lewis and Polacheck (1958b[120]) reported sequences of the onset of calcification in the cusps of the crowns of four permanent mandibular teeth (premolars, first and second molars). Calcification was first noted in the first molar; the second molar was usually the last of these teeth to calcify. The second premolar usually preceded the second molar in order of calcification, but there was considerable variability, and, in many children, priority of calcification within this pair of teeth could not be determined.

Sex differences in timing and variability of timing were smaller for dental maturation than for skeletal maturation, despite the fact that Garn and his colleagues documented considerably more variability in timing than had been reported previously (Garn *et al.*, 1958a[119], 1958b[120]; Garn, Lewis & Polacheck, 1959[121]; Garn, Lewis & Bonné, 1962a[103]; Lewis & Garn, 1960). Combining data from both sexes for five posterior mandibular teeth, Lewis and Garn (1960) showed that variability of timing increased with age (Table 5.2). Variability, adjusted for the period since conception, differed between stages of dental maturation, but this variability was less than that of skeletal age or dental emergence (Fig. 5.10). The median ages at which maturational stages were reached for the third mandibular molar tooth were no more variable than those for the other posterior mandibular teeth when corrected for the mean ages by calculating coefficients of variation (Table 5.2).

The rate of maturation of the third molar tooth showed marked constancy within individuals. If an early stage was advanced in timing, subsequent stages tended to be advanced also. The rate of maturation of this molar tooth was not correlated with age at menarche, but there were low correlations (0.1 to 0.4) with age at epiphyseal fusion at the proximal end of the tibia, and the age at completion of epiphyseal fusion in the hand–wrist (Garn *et al.*, 1962b[126]). These correlations were much lower than the corresponding correlations for the second mandibular molar tooth. In general, rates of dental maturation were only moderately associated with the timing of pubescence (Garn *et al.*, 1965b[108]). It was shown that dental maturation was less influenced by hormones than was skeletal maturation and that skeletal and dental maturity were essentially independent within chronological age groups (Garn *et al.*, 1965b[108]).

The timing of dental maturity stages were only moderately inter-

Table 5.2. *Variability* (*years*) *of timing of maturation for posterior mandibular teeth* (*sexes combined; data from Lewis & Garn, 1960*)

| Stages | Percentiles | |
|---|---|---|
| | 5 | 95 |
| *Beginning calcification* | | |
| First premolar | 1.6 | 3.0 |
| Second premolar | 2.7 | 0.24 |
| Second molar | 2.8 | 4.8 |
| Third molar | 7.5 | 10.9 |
| *Crown completion* | | |
| First premolar | 6.0 | 8.1 |
| Second premolar | 6.7 | 9.3 |
| First molar | 3.1 | 4.9 |
| Second molar | 7.3 | 10.2 |
| Third molar | 12.0 | 17.1 |
| *Apical closure* | | |
| First premolar | 11.2 | 14.0 |
| Second premolar | 12.1 | 15.4 |
| First molar | 8.8 | 11.6 |
| Second molar | 12.8 | 17.6 |
| Third molar | < 18.0 | < 26.0 |

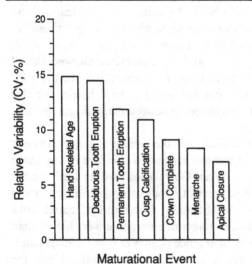

Fig. 5.10 A representation of the relative variability in the timing of selected aspects of skeletal, dental and sexual maturation. (Data from Lewis & Garn, 1960.)

correlated; reversals of sequence may occur during the maturation of pairs of teeth (Garn & Koski, 1957[96]). Garn *et al.* (1958b[120]) assigned one of five grades to each mandibular premolar and molar tooth and emphasized sequence variations between the second premolar and the second molar teeth. The proportion of children with a particular sequence varied from stage to stage of dental maturation. The order 'second premolar then first molar' was present for onset of calcification in 35% of the group, for the beginning of root formation in 58%, for apical closure in 7%, for alveolar emergence in 9%, and for reaching the occlusal plane in 9%. Almost all those with a 'second premolar then second molar' sequence of tooth formation had a corresponding sequence of emergence, but only half of those with a 'second molar then second premolar' sequence of tooth formation had the same sequence of emergence (Garn & Lewis, 1957[98]). The 'second molar then second premolar' sequence of tooth formation was associated with rapid maturation of the second molar.

Reference data for stages of dental maturation, based partly on work done at Fels, were reported by Garn *et al.* (1958a[119], 1959[121], 1960c[123], 1967f[147]). Fels data did not show secular trends in the rates of dental maturation, but girls tended to be advanced over boys by about 0.3 years (Garn *et al.*, 1958a[119]). The sex differences, expressed as percentages of the mean ages of occurrence, were about the same for calcification and emergence. These percentage differences were about 3%, but those for skeletal maturation were considerably larger (knee, 19%; hand–wrist, 12%). The data provided by Garn and his associates allowed dental maturity to be assessed from birth to 14 years. Garn recommended a 45° oblique radiograph of the jaw, obtained using a head-holder for this purpose. Since then, other methods based on panoramic radiographs have been developed.

It was reported that the correlations for the timing of stages between teeth were about 0.4; these correlations tended to decrease as the mean intervals between stages became longer. Correlations within teeth tended to be greater than correlations between teeth, indicating some autonomy of individual teeth in their rates of maturation. Furthermore, correlations that related stages of calcification or stages of emergence were higher than those that related calcification to emergence (Lewis & Garn, 1960). Later, Wolánski (1966a, 1967) devised a complex scoring method for the maturation and emergence of teeth that was based on the time a tooth remained in the same stage. With this method, the sum of the scores for all the teeth was used to assign a dental age.

Garn, Lewis & Bonné (1961a[102]) reported that the maturation of the mandibular premolar and molar teeth was delayed when the third molar tooth failed to form (agenesis). Also, children with third molar teeth, in the same families as those in whom this tooth did not form, tended to be delayed in the maturation of the premolar and molar teeth. They concluded that absence of the third molar tooth expressed factors that delayed dental maturation between birth and 8 years.

In an extension from this work, Garn *et al.* (1961a[102]) compared the timing of dental maturation in three groups: (i) those with agenesis of the mandibular third molar, (ii) siblings of those in group (i) in whom the third molar was present, and (iii) unaffected siblings from unaffected sibships. Cusp calcification and emergence in the other posterior mandibular teeth tended to be delayed in those with third molar agenesis, and there were small delays in their siblings. The siblings may have had the same genotype as the affected individuals, but differed in its expression, or the siblings may have had only part of the genetic constitution that leads to third molar agenesis. Third molar agenesis may represent the extreme expression of a variable factor affecting the rate of dental maturation. Children with third molar agenesis tended to have rapid maturation of the second premolar, relative to the second molar.

Fels scientists, led by Garn, have shown clearly that there is genetic control over various stages of dental maturation and emergence and that sequences of maturation and emergence between pairs of teeth are, at least in part, genetically determined (Garn, 1961c[59]; Garn *et al.*, 1956a[124]; Sontag & Garn, 1957[286]; Lewis & Garn, 1960).

### Sexual maturation

In two outstanding papers, Reynolds and Wines (1948, 1951) described the maturation of secondary sex characters in girls and boys and published 'standards' to assist grading, together with plots of the distributions of ages at which the grades were first noted. An example for girls is given in Fig. 5.11 which shows the cumulative percentages for the second grades of maturation of the breasts and of pubic hair.

Reynolds and Wines (1948) reported that the timing of early grades in the maturation of the breasts was more closely related to age at menarche (r = 0.9) than was the early development of pubic hair (r = 0.7). There were high correlations (r = 0.9) between the ages at which the early and later grades of sexual maturity were reached in girls, but breast size and breast shape were not related to the timing of breast maturation.

Genital maturation began before the appearance of pubic hair in 70% of boys, but growth in stature and weight accelerated when pubic hair was at grade 2 (Reynolds & Wines, 1951). The timing of the maturation of pubic hair was highly correlated between grades with coefficients ranging from

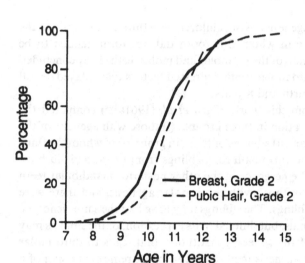

Fig. 5.11 Cumulative percentages for the prevalence of grade 2 of pubic hair and breast maturation. (Reynolds, E.L. & Wines, J.V. (1948). Individual differences in physical changes associated with adolescence in girls. *American Journal of Diseases of Children*, **75**, 329–50. Redrawn with permission from American Medical Association, copyright 1948.)

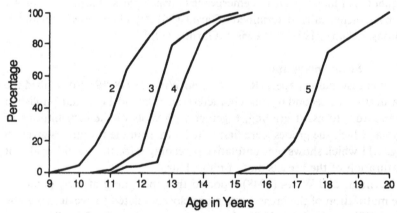

Fig. 5.12 Cumulative percentage prevalences for grades of maturation of the penis and scrotum. (Reynolds, E.L. & Wines, J.V. (1951). Physical changes associated with adolescence in boys. *American Journal of Diseases of Children*, **82**, 529–47. Redrawn with permission from American Medical Association, copyright 1951.)

0.5 to 0.9, despite marked variations in the duration of grades. For example, the mean duration was 0.6 years for grade 2, but 3.9 years for grade 4. Boys who were early in the attainment of grade 2 of pubic hair maturation tended to have profuse body hair at 18 years (Reynolds & Wines, 1951). Additionally, boys who achieved grade 2 of pubic hair or scrotal maturity early tended to have long intervals from then until the attainment of grade 5 (Fig. 5.12). Additionally, Reynolds and Wines (1951) reported that ectomorphic boys tended to be advanced in sexual maturation. These boys had long and slender limbs and trunk.

Reynolds (1951) noted familial resemblances in the distribution of pubic hair and other body hair in boys and documented the distribution of particular types of pubic hair distribution in both men and women. In men, a 'masculine' pattern was more likely to be associated with thigh hair, but not with chest hair. Omitting cranial and facial hair, the sequence of development of hair was: pubic, axillary, leg, thigh, forearm, abdomen, buttocks, chest, lower back, arm and shoulder. The development of hair on the upper body of men was unusual at 20 years, but was present in about 30% at older ages. Also, hair on the abdomen became more common with age in males, increasing from 14% at 14 years to 75% at 18 years.

### Weight at menarche

In an early study, Reynolds and Asakawa (1948) found that the first menstrual period (menarche) occurred earlier in girls with high values for either weight or weight/stature$^3$ at 8 years. Later, Garn and Rohmann (1966a[141]) reported that age at menarche was associated with fatness. Subsequently, others suggested that menarche occurred when a critical level was reached for weight. An analysis of several sets of data, including some from Fels participants, showed that weight at menarche varied from 28 kg to 79 kg (Johnston *et al.*, 1975). Consequently, the concept of a critical weight at menarche is not applicable to individuals and should be discarded.

### Other aspects of maturation

An investigation by Robinow, Johnson and Anderson (1943a) concerned angles measured on lateral radiographs of the foot during weight bearing. These angles were stable across age, with year-to-year correlations of 0.8. The major lateral antero-posterior angle changed little from 3 to 11 years despite large changes in footprints during this age range. Although this angle flattened with weight bearing, it did not flatten further when all the weight was carried on one foot. The antero-posterior angle was correlated between siblings ($r=0.4$), but it was not significantly correlated with weight/stature, severity of knock-knees, or pronation of the foot. Values of this angle were distributed almost normally, indicating that low arches are part of the normal population variation.

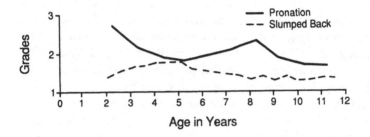

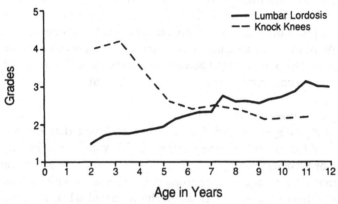

Fig. 5.13 The mean grades of elements of posture plotted against age (data from Robinow *et al.*, 1943b).

In a companion study, Robinow, Leonard and Anderson (1943b) reported a quantitative method for the analysis of posture in children. They prepared grades for body regions and provided photographs to assist the assignment of these grades. These photographs of front, back and lateral views were taken with the participants in a natural erect posture except that, for the lateral view, the participant was sometimes asked to move the arms slightly forward so that the curvature of the back could be observed. Robinow and his colleagues (1943b) considered these grades were applicable from 3 to 12 years and that they could be used directly without photographs. Correlations between raters were good for knock-knees and pronation, but not as good for the other aspects of posture.

Grades can be assigned easily for knock-knees using the rear view, if the feet are positioned with the medial borders parallel and the medial malleoli barely touching so that they do not support the ankles. This grading was based on the angulation of the Achilles tendon, tilting of the heel, and the relative position of the medial and lateral malleoli. Hyperextended knees and lumbar lordosis were graded also. Finally, they rated 'slumped back,' which was mainly due to a relaxation of the upper trunk and was common in preschool children. This condition was associated with kyphosis of the thoracic spine; in some young children almost the whole back was involved. The age trends in these grades are shown in Fig. 5.13. The average child had considerable knock-knees at 3 years and lumbar lordosis was common at about 10 years. The grades given to knock-knees were correlated with weight/stature ($r = 0.6$) at 3 and 10 years, but these variables were not correlated at 6 years. The grades of pronation were correlated with knock-knee grades in various age groups ($r = 0.2$ to $0.6$), but not with age or weight/stature. Robinow and his associates considered that inadequate muscle function was involved in most of these aspects of defective posture.

While not clearly an aspect of either maturation or development, Kingsley and Reynolds (1949) reported the relationship between illness patterns from birth to 5 years and birth order. First-born children had the highest mean incidence of gastro-intestinal upsets, feeding disorders, skin disorders, asthma and allergies, but second-born children led in respiratory and ear infections, tonsillitis, whooping cough, diarrhea, accidents and enuresis. These differences may have been related to the development of immune competence or to environmental variations.

**The non-invasive assessment of maturity**
It is not always practical to grade the maturity of secondary sex characters, such as the development of pubic hair, because this involves undressing and invasion of privacy. Furthermore, the grading of secondary sex characters is useful only during the age range when this discriminates among individuals. Similarly, the timing of peak height velocity or menarche is useful only during restricted age ranges. Assessments of skeletal maturity or dental maturity are based on radiographs and, consequently, there is some irradiation and high financial cost, in addition to which it is difficult to obtain assessments by trained individuals. To overcome these limitations, a method has been developed for the non-invasive estimation of maturity (Roche, Tyleshevski & Rogers, 1983c[269]).

To develop this non-invasive method, Fels data for stature, ages at peak height velocity and menarche, and skeletal age were analyzed (Roche *et al.*, 1983c[269]). The ratio (present stature/adult stature) is a measure of maturity because it has the same value (1.0) in all adults, and it increases monotonically with age during childhood. It had been assumed that this measure could only be obtained retrospectively when adult stature was known, but Roche *et al.* (1983c[269]) described how this ratio could be applied prospectively without the use of invasive procedures. The key to this was the use of adult statures that were predicted without the use of skeletal age (Wainer *et al.*, 1978). The ratio (present stature/predicted adult stature) was moderately but significantly correlated with age at peak height velocity and skeletal age after about 4 years.

To apply this method, the stature and weight of the child and the stature of each parent must be known. Roche *et al.* (1983c[269]) provided reference data for this ratio; as expected, the means for girls tended to exceed those for boys by amounts that increased with age. The standard deviations also increased with age until pubescence, after which they decreased rapidly. This non-invasive method of assessing maturity is recommended for many situations, but it should not replace the assessment of skeletal maturity from radiographs and the grading of secondary sex characters when there is particular concern about individuals.

### The regulation of maturation and its associations
#### Genetic regulation
Similarities within families in the sequence of ossification have provided evidence of genetic control over skeletal maturation (Reynolds, 1943; Sontag & Lipford, 1943[288]; Reynolds & Sontag, 1944; Garn & Shamir, 1958[164]; Garn, 1961c[59]; Garn & Rohmann, 1962c[138], 1966a[141], 1966b[142]; Garn *et al.*, 1961b[144]). Garn postulated that these familial tendencies could account for many of the differences in the relative lengths of body segments among adults.

In one study, parent–child correlations for the number of hand–wrist centers at an age were about 0.3 (Garn & Rohmann, 1962c[138]). The centers with the high parent–child correlations tended to be those with low communality indices which indicated that centers for which the timing was highly influenced by genetic factors were effective as predictors of timing for other centers. Garn, Rohmann and Davis (1963c[148]) showed that, as the average proportion of genes in common increased, there was a progressive similarity between pairs of related participants in the number of ossified hand–wrist centers at particular ages. Garn and his colleagues (Garn & Rohmann, 1962b[137]; Garn et al., 1967e[154]; Garn, Rohmann & Hertzog, 1969a[149]) also reported suggestive evidence of X-linked influence on the timing of ossification in the hand–wrist.

Genetic control over the rate of dental maturation was analyzed by Garn *et al.* (1961a[102]). The correlations for the rate of dental maturation were about 0.9 in monozygotic twins and about 0.3 between siblings. While teeth differed little in this regard, the pattern of correlations for various types of sibling pairs led Garn and his colleagues to suggest there may be X-linked influences on the rate of dental maturation. This conclusion may be incorrect because the ascending order of correlations in their data was brother–brother, brother–sister, sister–sister, but with X-linked inherit-ance the brother–brother and brother–sister correlations would be equal.

The body build of parents was related to the rate of maturation of their offspring. Skeletal maturation in the pre-school period and at 11 years tended to be advanced in children of parents with wide chests (Garn *et al.*, 1960b[122]). The basis for this familial association is unclear.

### Secular changes

Secular changes in the rates of maturation would imply a response to environmental alterations. The rate of maturation has accelerated in association with the well-documented secular increases in rates of growth and in adult size during the past century, particularly in Western Europe and North America (Roche & Davila, 1972[222]; Roche, 1979a[187]). The increases in adult stature imply that the secular increases in rates of maturation have been less than those in rates of growth.

More direct evidence was considered by Roche *et al.* (1974c[260], 1976a[261]) when they analyzed skeletal maturity data from US national surveys and compared the findings with reports from other studies including Fels. Convincing evidence of a secular trend was lacking, but data from high socioeconomic groups suggested maturation has become more rapid. In Fels data, evidence of secular changes in the timing of onset of ossification for the bones of the hand–wrist is lacking and there are only small differences in skeletal ages within parent–offspring pairs of the same sex when both were assessed at the same age (Garn & Rohmann, 1959[133], 1960b[135]; Roche *et al.*, 1975d[274]).

### Seasonal influences

The appearance of ossification centers within 6-month intervals was related to seasons by Reynolds and Sontag (1944) who analyzed data from 1 through 5 years. In each sex, most centers appeared in the interval February to August and fewest in the interval August to February, but these seasonal variations were small.

### Illness experience

Pyle and Sontag (1943) showed that children with more illnesses tended to mature rapidly, but illnesses were not associated with the sequence of ossification. Himes (1978a), in a review of the effects of nutrition on the rate of skeletal maturation, noted the possible confounding influences of illnesses and emphasized the variability present in normal children as reported by Garn and his associates (Garn & Rohmann, 1960a[134]; Garn et al., 1961b[144], 1966b[142]).

### Body composition and nutritional status

The possible associations between body composition and nutritional status with the rate of maturation have received the attention of Fels scientists over a long period. Sontag and Wines (1947[308]) did not find associations between the number of ossification centers at specific ages and the protein intakes of their mothers during pregnancy, but none of the mothers were suffering from a protein deficiency. There were, however, definite associations between rate of maturation and body composition in children.

Associations between radiographic measurements of calf tissues in girls and the timing of menarche and of the maturation of secondary sex characters were investigated by Reynolds (1946a). At ages 7 to 13 years, rapidly maturing girls had larger values than other girls for bone, adipose tissue and muscle widths and larger relative increases in these widths, especially muscle widths. The correlations between these tissue thicknesses and the timing of maturational events were about −0.5. Reynolds (1946a), in a novel approach, grouped girls with complete serial data who were matched for the total thickness of the calf at 7.5 years. He then chose contrast subgroups in which pubescent changes occurred either early or late. From 7.5 to 10.5 years, total calf thickness increased more in the rapidly maturing girls than in the slowly maturing girls, due to differences in muscle thickness. Additionally, Reynolds reported calf tissue thicknesses relative to the ages at which breast maturation reached grade 2. Rapidly and slowly maturing girls did not differ in bone widths at any age or in total calf width at 3 to 4 years before grade 2 of breast maturation was reached, but the total calf widths and muscle widths were larger in the rapidly maturing girls at older ages while the adipose tissue thicknesses were larger in the rapidly maturing girls at all ages.

Rapidly maturing children tended to have larger subcutaneous adipose tissue thicknesses and be taller and heavier, but the correlations between these variables and the timing of maturation were low, although they were slightly greater in girls than boys (Garn & Haskell, 1959a[86], 1960[88]; Garn, 1960b[55]). There were, however, close relationships between skeletal age and the percent of the body weight that was fat (%BF) in boys and the total weight of fat (TBF) in each sex (Chumlea et al., 1981a[24]).

**The timing and sequence of adolescent events**
Some analyses of Fels data have addressed variations in the timing and sequence of adolescent events (Roche, 1973[177], 1974a[178]). Most of these events can be defined only if serial data are available. For example, the age at peak height velocity can be determined only if data are available for about 3-year periods before and after this event. In the past, judgments of ages at peak height velocity were based commonly on annual increments beginning at birthdays, although rates of growth do not change at birthdays. This can cause misinterpretations as shown earlier (see Fig. 2.5).

There are differences in the sequences of maturational events during adolescence that include peak rates of bone elongation and of increase in bone breadths, peak height velocity, menarche and age at ossification of the ulnar sesamoid. Similar variations occur, as noted earlier, in the sequence of grades of skeletal and dental maturity.

**The development of hearing ability**
Beginning in 1975, a major study of hearing ability in relation to noise was commenced at Fels (Roche *et al.*, 1977b[266], 1978[267], 1983d[220], 1983e[259]). The aim was to provide knowledge that would assist public policy decisions concerning noise and help to explain why the sexes were similar in hearing ability before pubescence, but hearing ability was much better in girls than boys by 18 years. This work was supported financially by the Environmental Protection Agency and by the Aerospace Medical Research Laboratory at Wright–Patterson Air Force Base. It was terminated after 5 years because the Federal Government considered noise was not a problem. In this study, historical data were obtained for lifetime and recent exposure to noise, health histories were recorded, and otological inspections were made. The noise exposure data were used to obtain (i) a total 6-month quantitative score for each participant, and (ii) the number of individual events associated with noise exposure for each participant. One aim was to relate changes in hearing ability to growth and maturity.

Hearing ability (auditory thresholds) was measured in an audiometric booth located in a very quiet area. At each age, girls had better and less variable hearing ability than boys and, in each sex, hearing ability increased slightly with age (Roche *et al.*, 1978[267], 1979b[248]). Some of this age trend could have been due to greater concentration on the test by older children, better fit of the earphones to older children, and increasing familiarity with the procedure ('examination effect') as age and the number of examinations increased. 'Examination effects' were present in the data which lead to artefactual 'improvements' with repeated testing (Roche *et al.*, 1983a[255]). This may help explain why mean hearing ability in the Fels

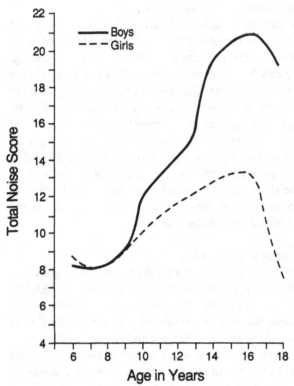

Fig. 5.14 Total noise scores for boys and girls at ages from 6 to 18 years (data from Roche *et al.*, 1982b[219]).

children was slightly better than the US national reference data at low frequencies, and markedly better than these reference data at high frequencies (Roche *et al.*, 1979b[248]; Roche, Chumlea & Siervogel, 1982b[219]).

Noise exposure from questionnaire data increased with age to 16 years, particularly in boys (Fig. 5.14). Noise exposure was not significantly correlated with body size or level of maturity in boys, but more mature girls at 13–15 years, who were also taller and heavier, were exposed to more noise. Despite this, the Fels data suggested that rapid maturation and tallness, especially in girls just before menarche, was associated with better hearing. Exposure to firearms, loud stereo music, loud TV, and farm machinery tended to be associated with reduced hearing ability, as were illnesses, abnormalities of the auditory meatus (ear canal), and of the color of the tympanum (ear drum, Roche *et al.*, 1982b[219]; Siervogel *et al.*, 1982d). Systolic and diastolic blood pressures and iris pigmentation were not

Table 5.3. *Descriptive statistics for* $L_{eq(t)}(dB)$[a] *in sound source categories with significant* ($p < 0.05$) *sex effects* (*data from Roche et al., 1982b*[219])

| Categories | | Mean | SD |
|---|---|---|---|
| Home, radio, TV | Boys | 74.1 | 7.4 |
| | Girls | 70.8 | 7.7 |
| Sleep | Boys | 58.3 | 6.4 |
| | Girls | 55.1 | 8.3 |
| School, normal class | Boys | 74.5 | 5.7 |
| | Girls | 68.9 | 6.6 |
| To and from school | Boys | 79.1 | 6.2 |
| | Girls | 69.1 | 8.2 |
| School year vs summer | | | |
| Sleep | School year | 56.2 | 7.8 |
| | Summer | 61.2 | 2.2 |
| Live music | School year | 82.4 | 9.0 |
| | Summer | 90.7 | 6.0 |

[a] $L_{eq(t)}$ is the equivalent sound exposure (dB) for the period $t$ on a log scale.

associated with hearing ability (Roche *et al.*, 1978[267], 1979b[248], 1982b[219], 1983b[256]).

The final report from this study included an analysis of noise exposure using data from 24-hour measurements (Roche *et al.*, 1983d[220]). These data were obtained for 3-minute intervals within each day and they were combined with a detailed activity diary. The highest levels of noise came from lawn mowers, boats, live music, riding a school bus, and recesses or assemblies at school. Boys were exposed to significantly more noise than girls for many noise sources (Table 5.3). The amount of noise associated with specific activities decreased with age for conversation, outdoor activities, and school activities, but increased for live music. The duration of exposure also decreased with age except for music and noise associated with automobile use. The amounts of noise exposure during the school year and the summer did not differ significantly in the Fels group (Siervogel *et al.*, 1982a).

### Developmental changes in skin reflectance
Studies of skin reflectance at ages 1 to 90 years were reported by Garn and his colleagues (Garn *et al.*, 1956c[162]; Garn & French, 1963[81]). Measurements were made on areas exposed to light, on protected areas and on the areola and scrotum where reflectance is markedly dependent on stimulation by sex hormones. Melanin is the major pigment involved in skin reflectance at the wave length used by Garn.

All skin sites reflected more light in infancy; this difference was marked for the areola and scrotum. Towards the end of pubescence, a large sex difference developed in the pigmentation of the chest and there was a reduction in reflectance from the areola, forehead and arm sites.

In men, the medial aspect of the upper arm and the chest reflected most light and the scrotum and areola reflected least. In men, reflectance tended to increase with age at the arm, chest and areolar sites, but decreased at the forehead site. In females, there was a marked decrease in areolar reflectance during pubescence, adulthood, and pregnancy, but pregnant women did not differ from controls in reflectance at the arm, breast or forehead sites (Garn & French, 1963[81]). Women who had delivered 3 to 11 months previously had less areolar reflectance than age-matched women of the same parity who had not been pregnant recently.

Work on the methodology of assessing maturity is, at least temporarily, in abeyance but analyses continue that relate established measures of maturity to the other recorded data.

# 6    Bones and teeth

'Can these bones live?'

*Ezekiel* 37:3 (*ca.* 630–570 *BC*)

Most of this chapter will be concerned with bones and teeth which share a common embryological origin. The Fels studies of these organs will be grouped as follows: (i) skeletal growth, (ii) growth of specific skeletal regions, (iii) skeletal variations, (iv) skeletal mass, and (v) teeth.

### Skeletal growth

Studies of skeletal growth at Fels have attempted to determine the amounts of growth at specific locations and to examine changes in overall dimensions such as length and width. In an early attempt to establish a natural bone marker for use in studies of bone elongation, based on serial radiographs, Pyle (1939) observed the nutrient foramen of the radius. She found the groove leading to the foramen made it difficult to locate a fixed radiographic point, although the foramen could be recognized in about 90% of radiographs of children. Some difficulties were associated with the presence of multiple nutrient foramina in about 20% of radii and the changing location of points at the external end of the obliquely aligned nutrient canal as the cortex thickened. This interesting exploratory study did not establish the nutrient foramen as a suitable fixed point from which length measurements could be made.

Fels data have been analyzed to provide reference data derived for bone lengths and widths measured in serial radiographs. Garn *et al.* (1972[93]) presented such data for the bones of the hand at annual intervals from 2 through 18 years. These workers defended the use of Fels data on the basis that these data were derived from a well-nourished contemporary population and, therefore, they should be applicable fairly generally to US whites. All the length measurements included the epiphyses and, like the

157

widths, they were measured to the nearest 0.1 cm. These data, which showed a uniform tendency for the mean values to be larger for males than females until about 14 years but not at older ages, were used to make 'profile' plots of the standard deviation levels for the lengths of each hand bone in individuals (Poznanski *et al.*, 1972). These 'profile' plots are valuable in the recognition of abnormal patterns of hand-bone lengths and widths that may indicate the presence of a syndrome.

Garn and his colleagues (1972[93]) calculated the coefficient of variability (CV) for the lengths of each hand bone. These CV values were high in each sex for distal phalanges II and V and middle phalanx V, but variability was low for proximal phalanges II and IV in boys and for proximal phalanges III and V in girls. In general, bone lengths tended to be more variable in females than in males. Additionally, it was claimed that elongation tended to be proportional from infancy through adulthood with the addition of similar percentages to the lengths of all bones each year, despite the known changes in rates of growth at pubescence that had been established by others. In reviews, Roche (1978c[186], 1986b[202]) directed attention to these and other reference data from the Fels study and evaluated their utility.

Mean lengths of the tibia and the radius at ages from 1 month through 18 years were reported by Gindhart (1971, 1972, 1973). Until adolescence, there were only small sex differences in the length of the tibia but the radius was significantly longer in males than females at almost all ages. Additionally, Gindhart presented increments for four intervals during the first year, for 6-month intervals to 12 years, and then for annual intervals. These increments showed pubescent growth spurts at 12 to 15 years in males and 9 to 12 years in females for both the tibia and the radius. There were some significant differences between the sexes in tibial length increments but they varied in direction and timing and may have been due to chance, except for significantly larger increments in males from 12 to 17 years. There were few statistically significant sex differences in the increments in radial length before 10 years, but the increments were significantly larger in girls from 10.5 to 12 years and in boys from 13 to 18 years.

Gindhart analyzed 6-month increments in tibial and radial lengths from 1 to 9 years in relation to seasons and childhood disorders. After grouping the Fels participants by month of birth, she concluded that the tibia elongated most from October to December and least from April to June. In the male radius, the maximum rates of elongation occurred in October and November and the minimum rates occurred in April and May. The seasonal fluctuation was most marked in the male radius and least marked in the female tibia. The timing of maximum and minimum increments differed more between bones within each sex than between sexes for the same bone which was difficult to interpret on any nutritional or hormonal basis. Presumably, these differences in timing were related to systematic variations in end organ responsiveness.

Three-month increments did not show seasonal trends in the rates of elongation of the tibia in males during infancy, but the maximum increments for the male radius occurred from June to August and the minimum increments from February to April (Gindhart, 1971). The corresponding findings for the female radius were less clear, but the minimum increments tended to occur from February to April. Gindhart (1971) commented on the findings of Reynolds and Sontag (1944) concerning seasonal influences on the rate of growth in stature from 1 to 5 years. The timing of the small seasonal effects found in this earlier study was the opposite of that found by Gindhart (1971). In both sets of data, there was marked variability during the first year, with the girls being more variable than the boys in elongation of the tibia, but there was little sex difference in the rates of growth in stature and in the length of the radius. Gindhart did not find significant relationships between childhood disorders and the rates of elongation of the tibia or the radius in either sex.

Using skeletal dimensions that were mostly bone lengths, Roche (1974a[178]) calculated annual increments and estimated the ages at the maximum rates of growth for individual participants in the Fels study and the Child Research Council (Denver) Study. The skeletal dimensions were measured with high reliability on standardized radiographs. There was remarkably good agreement between data from the Fels and Denver studies. The differences between dimensions in the mean ages of their maximum increments were small but the variances were large. Nevertheless, they were similar to the variance for age at peak height velocity with the exception of the cranial base lengths that had larger variances. The order of the mean ages at which these spurts occurred differed by sex, and there was marked sequence variability within individuals. The maximum increments in metatarsal V, metacarpal II and sella–nasion (a cranial base length) tended to occur relatively earlier in the sequence in girls than in boys. Associations between long bones in the timing of their maximum increments tended to be higher for girls than for boys and were markedly higher for bones in the upper extremity than for other groups of bones. It was concluded that, although the cells at the sites of elongation for various bones (the epiphyseal zones of long bones) respond to the same hormones, these cells must differ among individuals in the levels of hormones to which they are sensitive.

In another approach to the study of bone elongation, Falkner and

Roche (1987) reported high correlations between prenatal femoral lengths and measurements of recumbent length or crown–rump length soon after birth. Regression analyses showed that, soon after birth, recumbent length or crown–rump length could be estimated from femoral length with errors of 0.1–0.3 mm. These findings led to the conclusion that the continuum of prenatal and postnatal skeletal elongation could be evaluated using prenatal femoral lengths from ultrasonic examinations and postnatal data for recumbent length or crown–rump length.

In some Fels publications, comments have been made about the possible retarding effects of undernutrition on bone elongation (Robinow, 1968; Falkner, 1975). Contrariwise, Garn, Greaney and Young (1956d[84]) showed that tibial length was greater in fat infants (highest quartile of calf adipose tissue thickness at 1 month) than in lean infants (lowest quartile) but these groups did not differ significantly in tibial length after infancy. This could be related to the decreases in the differences in fatness between these groups after infancy. Similarly, at birth the lengths of long bones and crown–rump length were correlated (r = 0.2) with adipose tissue thickness over the 10th rib (Garn, 1958a[52]).

Reynolds (1944, 1946a) found marked increases in the total breadth of the tibia from birth to 6 months, after which there was a deceleration until the pubescent spurt. This spurt began at the time of peak height velocity in boys, but earlier than this in girls. In each sex, the spurt in width lasted about 2 years, being longer in boys than in girls. Calf bone widths, relative to the total width of the calf, were larger in males than females from birth to early adulthood (Reynolds, 1944, 1948; Reynolds & Grote, 1948).

Hertzog et al. (1969) reported correlations of 0.8 between tibial length and stature and that stature could be predicted from tibial length with an error of 3 cm. The systematic over-prediction in older individuals allowed estimations of the decrease in stature with aging. These decreases occur in the trunk, not the limbs. The estimated losses in stature with aging of about 3 cm for men and 6 cm for women were in agreement with those from serial data.

### Growth of specific skeletal regions
#### Cranial vault

Using standardized lateral radiographs, a large amount of research has concerned the size and shape of the cranial vault. Cranial shape has been studied by physical anthropologists for more than a century because it reflects brain shape and because cranial shape was used to discriminate between racial groups. Using standardized lateral radiographs, Young (1956) investigated reported methods for the description of cranial shape.

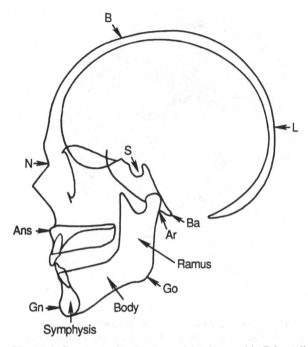

Fig. 6.1 A diagram to show some cranial points used in Fels studies. N = nasion; B = bregma; L = lambda; Ba = basion; S = sella; Ar = articulare; Go = gonion; Gn = gnathion; and Ans = anterior nasal spine.

Many descriptions have been subjective but others have been based on various craniometric procedures. Young's major aim was to compare the latter and determine which was the most useful. The measures compared were (i) the angle lambda–basion–nasion, (ii) the ratios chord length/arc length, height of arc from chord/chord length, and the radius of curvature for specific cranial bones, and (iii) for each bone, the lengths of nine equally spaced perpendiculars from each chord to the corresponding arc, adjusted for chord length (Fig. 6.1). All except the angle lambda–basion–nasion provide bone-specific data. Application of these methods consistently showed the frontal bone was more curved than the parietal bone and that frontal curvature was greater in females than in males. The inter-correlations between values from these methods exceeded 0.9, showing that they all yielded practically the same information.

Analyses of changes during growth in the size and shape of the frontal and parietal bones in males were made by Young (1957). He measured the ectocranial arcs and chords of these bones on lateral radiographs and showed that the patterns of change in size and shape were similar for these

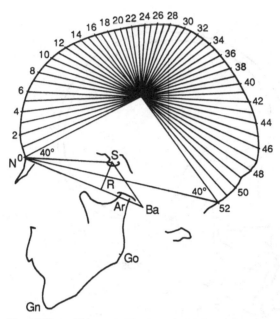

Fig. 6.2 The radii measured by Lestrel and Brown (1976) in their analyses of cranial vault shape. R is the midpoint of a line from sella (S) drawn perpendicular to nasion–basion (N–Ba). The line N–R was extended posteriorly to the ectocranial margin. From this point and from N, lines were drawn at 40° to N–R (extended) until they intersected. Fifty-two radii were drawn to the ectocranial margin from the intersection of these lines (centroid). (Lestrel, P.E. & Brown, H.D. (1976). Fourier analysis of adolescent growth of the cranial vault: A longitudinal study. *Human Biology*, **48**:517–528. Redrawn with permission from *Human Biology*.) For other abbreviations see Fig. 6.1.

two bones, with rapid decelerations to 4 years after which there was little change. Indeed, from 4 to 16 years increases in size were not detectable in 50% of the boys. Growth appeared to be complete earlier in the parietal than in the frontal. The arc lengths of these bones appeared to be interrelated in a simple allometric fashion with low negative correlations between the bones. The curvature increased to 2 years for the parietal and to 3 years for the frontal, after which these bones progressively flattened. Young measured cranial thickness at nine sites in each bone; the means of these thicknesses increased to 16 years with some decelerations and with correlations of about 0.5 between the frontal and parietal.

Using landmarks on the external (ectocranial) surface of the cranium, Young (1957) measured the angles of the triangle lambda–bregma–nasion. The angle at bregma increased gradually to 16 years, while the angle at nasion decreased and there was little or no change in the angle at lambda.

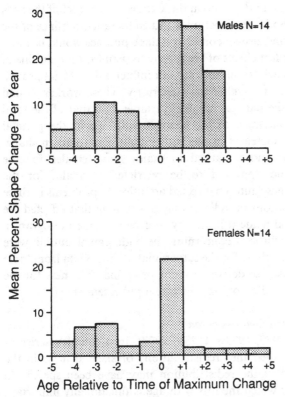

Fig. 6.3 Mean annual absolute percentage changes in cranial base shape plotted
against years in relation to the maximum change in shape. (Lestrel, P.E. &
Brown, H.D. (1976). Fourier analysis of adolescent growth of the cranial vault:
A longitudinal study. Human Biology, **48**: 517–528. Redrawn with permission
from *Human Biology*.)

These findings indicated more anteroposterior growth of the frontal bone
than of the parietal.

After a lapse of about 20 years, Lestrel and Brown (1976) returned to
this topic area. They obtained much more accurate descriptions of the size
and shape of the ectocranial margin by applying Fourier analysis to the
lengths of 52 radii drawn from a centroid (Fig. 6.2). Data from serial
radiographs for the age range from 4 to 18 years showed marked changes
in size at about the age of peak height velocity that were larger in males
than in females (Fig. 6.3). The data in this figure are for only the first four
Fourier terms; these accounted for about 96% of the variance. These
workers also found a small tentative 'mid-growth spurt' at 10 to 11 years
in males and at 8 to 9 years in females.

To help define the role of the brain in these changes, a subset of the data were used to analyze both the ectocranial and endocranial outlines of the parietal bone. Lestrel and Brown considered these outlines would be more likely than others to reflect effects of brain growth than the total outline of the vault because the parietal outlines are not influenced by facial growth anteriorly or by muscular attachments posteriorly. These workers found that the curvature of the parietal was less variable than that of the total cranial outline and that most of the changes with age in the parietal curvature were due to increases in size rather than alterations in shape. Pubescent spurts were more marked in males than females for the ectocranial margin and appeared to be restricted to males for the endocranial margin. These authors directed attention to problems in these data due to slight variations in radiographic positioning that affected the endocranial outline and made it necessary to exclude some radiographs. Consequently, the conclusions concerning the endocranial outline were less well established than those for the ectocranial outline. Data from these paired outlines were used to derive an index of cranial thickness for the parietal bone; this index did not demonstrate a pubescent spurt.

### Cranial base and facial skeleton

Israel (1967a, 1968a, 1970, 1973a, 1973b, 1977), in a complex series of studies of adults, described the 'redistribution' of skeletal mass in the craniofacial area. Part of the redistribution may have been specific to tooth-bearing bone for which the loss with age is functionally important when it is due to periodontal disease and leads to reduced support for teeth and their ultimate loss. A second type of bone loss affected the remainder of the craniofacial skeleton where the changes resembled the remodelling and redistribution that occurs in the second metacarpal (Garn *et al.*, 1967e[154], 1967g[155]).

Israel (1967b) restricted his Fels sample to those with teeth in the area of interest and with pairs of lateral head radiographs taken at least 13 years apart. In this group, the loss of tooth-bearing bone was similar in each sex but the amount was small, presumably because there was no loss of teeth. 'Basal' bone, in the mandible, increased with age due to continued tooth migration towards the occlusal plane and possible apposition on the inferior border of the mandible where cortical thickness increased. There was little association between these mandibular changes and corresponding changes in the remainder of the craniofacial skeleton or the second metacarpal. The cortical thickness of the mandible was not correlated with that of the second metacarpal and there were only low correlations between cranial thickness and the width of the second metacarpal. In these adults, there were increases in cranial thickness, total ectocranial length, upper face height and cranial base length, indicating that, unlike most parts of the skeleton, apposition exceeded resorption in these areas during middle age.

Israel (1971, 1973a, 1973b, 1977) found a generalized expansion of about 5% in the craniofacial skeleton after middle age, including ectocranial and endocranial dimensions. There were considerably larger increases in the thickness of the frontal bone, and the size of the paranasal sinuses and the sella turcica. In women, the increase in cranial thickness differed by area; it was about 1% at lambda but about 11% at bregma which is placed more anteriorly. These points are shown in Fig. 6.1. The increases in cranial thickness during intervals of 13 to 28 years in women were significantly different from zero at points anterior to bregma but not at points posterior to this craniometric landmark. In women, the external dimensions of the cranium increased until about 70 years due to apposition and there was concomitant resorption at the endocranial surface; after 70 years, resorption predominated at both the ectocranial and endocranial surfaces. Most mandibular dimensions increased about 4% in women during the same intervals but the height of the ramus increased only 1% and the change in the gonial angle was not significant. The gonial angle (Ar–Go–Gn) describes the relationship between the body and the ramus of the mandible. Similarly a change did not occur in the saddle angle (basion–sella–nasion). The saddle angle is often used in descriptions of the cranial base; sella is a constructed point in the middle of the pituitary fossa. Cranial base lengths (Ba–N, S–N, Ba–S; see Fig. 6.2) and upper face height (nasion–anterior nasal spine) also increased during adulthood by about 4%, and there was an increase in the anteroposterior thickness of the symphysis of the mandible of about 6% (Israel, 1971, 1973a, 1973b). The anterior nasal spine is on the anterior margin of the maxilla at the margin of the nasal aperture. The symphysis of the mandible is where the two halves of this bone join anteriorly. This extensive work by Israel is of particular importance because there are so few studies of changes in the craniofacial skeleton during adulthood.

Reference data for nine cranial base lengths and three angles from birth to 18 years, adjusted for radiographic enlargement, were reported by Ohtsuki *et al.* (1982a). Almost all these dimensions increased rapidly to 2 years, after which there were decelerations followed by pubescent spurts. The lengths were similar for the two sexes to 9 years, after which the values for the boys tended to be the larger. Cranial height and cranial length, both measured to the ectocranial margin, did not increase significantly after 7 years. A factor analysis of these data for 3-year age ranges showed the

Table 6.1. *Mean annual increments (mm/year) near the time of pubertal spurts (data from Roche & Lewis, 1976[250])*

| Length | Boys | | | Girls | | |
|---|---|---|---|---|---|---|
| | Before | During | After | Before | During | After |
| S–N | 0.2 | 1.5 | 0.9 | 0.2 | 1.2 | 0.4 |
| Ba–N | 0.5 | 2.5 | 1.5 | 0.6 | 2.0 | 0.8 |
| Ba–S | 0.3 | 1.8 | 0.6 | 0.3 | 1.4 | 0.4 |

segments of the cranial vault and base were represented in different factors for these age groups which suggested that the cranial vault and base had their own growth patterns (Ohtsuki *et al.*, 1982b). The changes with age in factor structure were consistent with the ages at which changes occurred in the growth rates of individual segments (Ohtsuki *et al.*, 1982b). Only two factors were extracted for the 0–3-year group that may be labelled (i) cranial vault size and cranial base length, and (ii) posterior cranial base length. This indicated a uniformity of growth rates for these groups of dimensions during infancy. Four factors were extracted for the older age groups of boys that were labelled (i) cranial vault size, (ii) posterior cranial base length, (iii) basisphenoid length, and (iv) presphenoid length. The patterns were generally similar for the girls except at 13–15 years when the differences noted may have reflected the earlier fusion of the spheno-occipital synchondrosis in the girls than in the boys.

The common occurrence of pubertal spurts in the cranial base and the mandible has been documented in a series of papers (Lewis & Roche, 1974; Roche & Lewis, 1974[249], 1976[250]; Lewis, Roche & Wagner, 1982, 1985). These studies were based on lateral head radiographs, taken at annual intervals, in which three cranial base lengths and three mandibular lengths were measured. All the measurements were shown to be highly reliable and they were corrected for the known enlargement that occurs with radi-ography. Spurts in cranial base lengths were recorded for a participant when an annual increment exceeded the immediately preceding increment by at least 0.75 mm in the boys and 0.5 mm in the girls. These sex-specific criteria were employed because the prepubertal growth rates were more rapid in boys than girls. The ages at the spurts were recorded as the mid-points of the annual intervals for which they were first observed. These spurts tended to occur later in boys than girls and they tended to be larger in boys (Table 6.1). These spurts were commonly asynchronous within individuals although synchrony would seem to be functionally desirable.

Relationships between the growth of the cranial base and mandible with stature were investigated by comparing groups of tall and short children. The total increments in the cranial base and mandible after peak height velocity were larger for short boys than for tall boys in all dimensions except Go–Gn. This measurement from the angle of the mandible to its most anterior point represents the length of the body of the mandible. In tall boys, the spurts tended to be large for S–N and small for Ba–S. These distances are lengths of the cranial base from sella (S; the midpoint of the pituitary fossa) to nasion (N; the anterior end of the cranial base) and basion (Ba; the posterior end of the cranial base). The specialist will recognize the limitations of the definitions given in this chapter to these and other craniometric points; the articles referenced provide precise definitions. Tall girls tended to exceed short girls in the size of their cranial base spurts. The sizes of the pubertal spurts were also analyzed in relation to rates of maturation. In rapidly maturing boys, spurts in the cranial base tended to be smaller for Ba–N, which is a measure of the overall length of the cranial base and they tended to be larger for S–N. In rapidly maturing girls, the spurts tended to be larger in Ba–N and Ba–S but not S–N.

In the boys, spurts were most common for Ba–N and for two mandibular lengths (Go–Gn and Ar–Gn). The latter is a measure of the overall length of the mandible. The highest prevalences in the girls were for cranial base lengths, particularly Ba–N. While the prevalence of cranial base spurts was similar in the two sexes, spurts in the mandible were much more common and larger in boys than in girls. In each sex, spurts were least common for Ba–S. Within individuals, there were usually decreases in the rates of growth before the spurts and only slow growth after the spurts. Corresponding patterns were more evident at the upper centiles than the lower ones and when the data were related to maturational events, particularly age at menarche in girls.

In girls, cranial base spurts tended to occur at chronological and skeletal ages near 11.5 years, with similar variability of timing in relation to both these ages. In boys, their timing was less variable relative to age at peak height velocity. For girls, the variability of timing in relation to menarche and ulnar sesamoid ossification was slightly less than that in relation to peak height velocity.

Spurts in the cranial base occurred about 1.6 years earlier in girls than in boys but their timing in relation to maturational events was similar in the two sexes. They occurred about 1 year after the onset of ulnar sesamoid ossification, but preceded peak height velocity by about 0.5 years and preceded menarche by about 1.5 years. In the boys, percentiles of increments for cranial base lengths decreased from 2–6 years before the

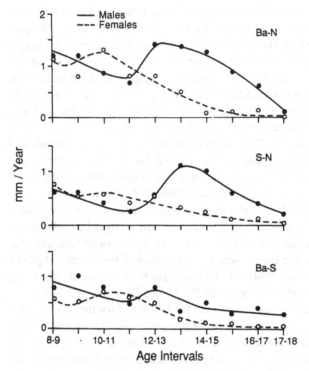

Fig. 6.4 Median rates of elongation of three cranial base lengths (Roche & Lewis, 1974[249]).

onset of ulnar sesamoid ossification. These decelerations were followed by accelerations from 2 years before to 2 years after the onset of ulnar sesamoid ossification for Ba–N and S–N. The acceleration in Ba–S ended, however, at about the time ossification began in the ulnar sesamoid.

Ages at spurts in pairs of lengths were consistently more highly correlated in girls (0.6) than in boys (0.4) but most of the correlations were significant in each sex. The median rates of elongation for cranial base lengths are shown in Fig. 6.4. In boys, Ba–N decelerated to 11.5 years and then accelerated to 13 years after which it again decelerated (Roche & Lewis, 1974[249]). Ba–N had a similar pattern of change in girls but the spurt occurred about 2 years earlier and was less marked. There were similar changes in S–N and Ba–S with more marked spurts for S–N in the boys but only small sex differences for Ba–S. The 90th percentiles for annual increments after 14 years were considerably higher in boys than girls for each cranial base length but there were only small sex differences at the 10th percentile level.

Table 6.2. *The mean total increments (mm) in cranial base lengths after selected ages and peak height velocity until 17.5 years (data from Roche & Lewis, 1976[250])*

| | After chronological ages (years) | | After peak height velocity |
|---|---|---|---|
| | 10 | 16 | |
| *Boys* | | | |
| Ba–N | 10.7 | 3.3 | 6.7 |
| S–N | 6.6 | 2.3 | 4.7 |
| Ba–S | 5.6 | 1.5 | 3.0 |
| *Girls* | | | |
| Ba–N | 7.4 | 2.4 | 5.4 |
| S–N | 4.7 | 1.9 | 3.7 |
| Ba–S | 3.2 | 0.4 | 2.1 |

Comparisons were made between Ba–N and Ar–Gn because of the similar lengths and orientations of these measurements that relate to the total cranial base and length of the body of the mandible respectively. There were smaller decelerations in Ba–N before the pubescent spurts and less gradual decelerations after them. Growth was slower in Ba–N than in Ar–Gn before, during, and after the spurts. Comparisons were also made between S–N and Go–Gn because of their similar orientations. S–N is a measurement of anterior cranial base length and Ar–Gn is a measurement of the length of the body of the mandible. The spurts were smaller in S–N and the decelerations after the spurts were also smaller. A similar comparison was made between Ba–S (posterior cranial base length) and Ar–Go (mandibular ramus height). There were smaller decelerations in Ba–S than in Ar–Go before the spurts and the spurts were smaller.

The total increments in cranial base lengths after peak height velocity and selected chronological ages tended to be larger in boys than in girls (Table 6.2). There was considerable elongation after 16 years except for Ba–S in girls (Roche & Lewis, 1974[249]; Lewis & Roche, 1988). Since the posterior part of the cranial base elongates until at least 17.5 years in boys and 16.5 years in girls, the brainstem probably elongates to corresponding ages. This elongation demonstrated that the distance from the foramen magnum to the pituitary fossa increased and, presumably, the length of the brainstem increased correspondingly. This is not surprising since cross-sectional data show increases in the weights and cholesterol content of the brainstem during pubescence.

To describe more fully these late changes in the growth of the cranial base, a double logistic model (sum of two logistic functions) was fitted to the serial data for individuals (Bock *et al.*, 1973; Roche *et al.*, 1977a[251]). This indicated that the pubescent phase of growth in the cranial base began at about 8 years in boys and 6 years in girls and that 95% of the adult lengths of cranial base dimensions was reached at about 15 years in boys and at 11.8 to 13.5 years in girls.

Sibling correlations for radiographic measurements of the cranium have been calculated using data for pairs of measurements made at the same ages (Byard *et al.*, 1984). These correlations were lowest during the first year of life and increased as adulthood approached. The patterns of change in these correlations varied during the intervening years. Most of the correlations were significant and many exceeded 0.5, which is the value expected for a completely heritable polygenic trait. This suggested that common sibling environment contributed to sibling resemblance for craniofacial dimensions.

Other studies of the cranial base have concerned the pituitary fossa and the sphenoid sinus. Analyses have been made of changes in the size of the pituitary fossa during childhood and adulthood. These studies are important because pituitary fossa size is related to the size of the pituitary gland. The Fels data were obtained from lateral cephalometric radiographs on which the outline of the fossa can be seen clearly. Silverman (1957) showed that measurements of the length and depth of the fossa were highly reliable in children and that they could be used to calculate its area. He presented selected percentiles of the area in relation to age and to stature. These areas showed a pubescent spurt in females but not males, that was particularly evident in the data for annual increments (Fig. 6.5). The relationship between the area of the pituitary fossa and stature was more nearly linear than the relationship with age. Relative to stature, the mean areas were greater in males than in females for the range of statures from 70 cm to 160 cm (which approximates the age range 1–13 years). At larger statures, the sexes did not differ in mean areas.

Other analyses showed that the pituitary fossa continued to enlarge during adulthood. This reflected bone loss, although aging in the craniofacial skeleton involves both remodelling and redistribution of bone (Israel, 1967a, 1968a, 1970). Israel (1970) measured radiographs of individuals aged more than 24 years and utilized serial data for some of these. Both the cross-sectional and the serial data showed an increase of about 10% in the area of the fossa during 14 years.

Attention has been directed to changes with age in radiographic measurements including the sphenoid sinus during childhood (Hinck & Hopkins, 1965). The interest in these studies is related to the light they shed

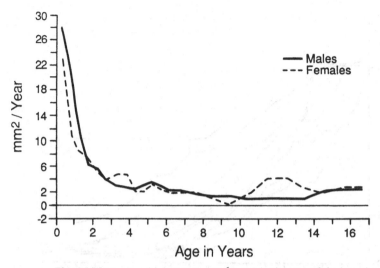

Fig. 6.5 Mean annual increments (mm²/year) in the area of the lateral outline of the pituitary fossa. (Silverman, F.N. (1957). Roentgen standards for size of the pituitary fossa from infancy through adolescence. *American Journal of Roentgenology, Radium Therapy and Nuclear Medicine*, **78**, 451–60. Redrawn with permission from Williams & Wilkins, Baltimore, MD.)

on the growth of the cranial base. The distance from the clivus (which is the part of the cranial base posterior to the pituitary fossa) to the sphenoid sinus showed near linear decreases in the Fels participants prior to pubescence (Fig. 6.6). These findings could, in part, have reflected changes on the dorsal surface of the clivus which, as will be described later, shifted anteriorly after the age of 9 years, but did not shift sufficiently to explain all the decreases noted by Hinck and Hopkins. The rate of decrease tended to be more rapid for boys than girls and it was not related to initial size. This distance was consistently larger in boys than girls to 13 years but not at older ages. There were only slight changes after pubescence.

### Cranial base shape
Serial studies have used Fels data to describe age changes in the shape of the cranial base. This topic is important biologically and in orthodontia and facial surgery but shape, in general, is difficult to describe and it is even more difficult to understand. This was expressed succinctly by Sherrington a few decades ago when he wrote: 'To record shape has been far easier than to understand it.' Leaving understanding to one side, it is extremely difficult to describe the shape of the cranial base in an accurate metric fashion.

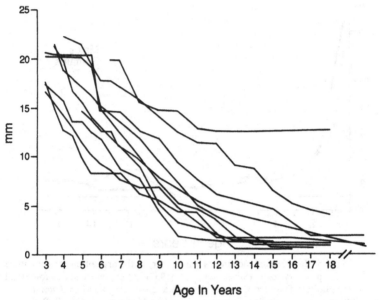

Fig. 6.6 Distances (mm) from the dorsal surface of the clivus to the sphenoid sinus in 10 girls. (Hinck, V.C. & Hopkins, C.E. (1965). Concerning growth of the sphenoid sinus. *Archives of Otolaryngology*, **82**, 62–6. Redrawn with permission from American Medical Association, copyright 1965.)

Changes in the relationships from 0.5 to 17 years between three cranial base planes defined by the foramen magnum, clivus, and planum sphenoidale and the plane of the palate were described by Koski (1961). In cross-sectional data, there were large changes in these relationships to 3 years but later changes were slight. There were, however, marked irregular changes in serial data for individuals in the angles between these planes during all the age ranges studied. Consequently, Koski rejected the concept of a fixed growth pattern for the craniofacial skeleton and the use of rigid norms for diagnosis, prognosis and classification.

The angle Ba–S–N, which is commonly called the saddle angle, is used as an inexact but useful index of the shape of the lateral silhouette of the cranial base. There has been considerable debate as to whether the saddle angle changes during growth. Cross-sectional data analyzed by Lewis and Roche (1977) and by Ohtsuki *et al.* (1982a) showed a significant linear decrease from 0 to 3 years, after which the changes were not significant to 40 years except for a decrease in males from 7 to 18 years. There were small consistent tendencies for the saddle angle to be larger in boys than in girls but there was a small sex difference in the opposite direction in adults.

Generalizing, the mean values were 139° at birth, 132° from 2 to 12 years, 131° from 12 to 24 years and 130° from then to 40 years.

Lewis and Roche (1977) reported that, after 2 years, the median increments in the angle Ba–S–N were negative for six age ranges that collectively extended from 2 to 30 years but the changes after 16 years were small. The high correlations between values 6 to 24 years apart attested to the reliability and stability of this measurement. Contrariwise, the correlations between successive increments within individuals were low, which reflected the small size of the changes. The saddle angle was correlated with Ba–N ($r = 0.3$) but the correlations with cranial vault length and stature were near zero. There were only minor differences in the saddle angle between groups with normal occlusion and those with malocclusion.

Age changes in the shape of the cranial base have also been described using Fourier analysis (Lestrel & Roche, 1984, 1986). Such an approach is needed because the irregular shape of the cranial base cannot be described effectively by a simple measurement such as the saddle angle. Additionally, Ba is the only conventional landmark on the endocranial surface of the base; S is a constructed point and N is ectocranial. The basic data for the Fourier analyses were the lengths of 61 radii measured on serial radiographs from a centroid that was chosen to reduce the effects of abrupt changes in curvature, especially near the posterior border of the pituitary fossa, Ba and the anterior part of the cranial base. This approach represented the endocranial outline accurately and allowed the irregular shapes to be measured while minimizing size differences. The dorsal aspect of the clivus and the anterior cranial base changed least with age while the pituitary fossa changed most. These data suggested that the cranial base was remodelled continually during growth.

The size of the cranial base was separated from its shape without ambiguity by making the area bounded by the Fourier construct equal between individuals. After adjustment for size, the mean shapes were similar for the two sexes in all age groups and these shapes changed with age within each sex. The data for females showed considerable changes between the shapes in infancy, pre-puberty and adolescence (Fig. 6.7). In each sex, the floor of the pituitary fossa tended to move superiorly from about 3 to 9 years and there was a marked antero-superior movement of the region near the tuberculum sellae, and anterior movement of the anterior wall of the pituitary fossa and the clivus near the dorsum sellae from 9 to 15 years. The shape of the cranial base did not change from adolescence to adulthood.

It was found that the region near the pituitary fossa was placed more

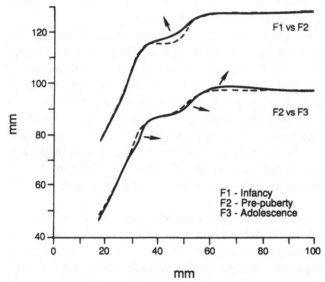

mm

Fig. 6.7 Mean coordinates of the shape of the cranial base in girls during infancy (F1), before puberty (F2), and in adolescence (F3). In each comparison, the interrupted line represents the earlier shape. (Lestrel, P.E. & Roche, A.F. (1986). Cranial base shape variation with age: A longitudinal study of shape using Fourier analysis. *Human Biology*, **58**: 527–540. Redrawn with permission from *Human Biology*.)

superior in females than in males during infancy. From then until pubescence, the region near the dorsum sellae was more posterior in females and the posterior part of the pituitary fossa was more inferior. In females during adolescence and adulthood, the tuberculum sellae, and the posterior part of the floor of the pituitary fossa were more inferior and the clivus near the dorsum sellae was more posterior. With a different superimposition, these changes could have been interpreted differently but, regardless of the superimposition used, it was shown that the cranial base changed in shape during growth and that there were sex differences in shape at an age.

This work by Lestrel provided important findings but it is more significant for the method that he introduced to measure the shapes of parts of the skeleton. Lestrel *et al.* (1991) illustrated the further application of this method in a study of nasal bone growth from birth to 1 year. Sex differences in shape remained after standardizing for size. These differences, which were mainly related to greater length and reduced thickness in females, were present from birth and became significant at 1 year.

Garn (1961a[57]) reviewed research opportunities and limitations related to malocclusion. He criticized conventional caliper measurements for the study of growth because they had been designed to compare races and not to describe 'the way the skull, the face, and the jaws actually grow.' Many caliper measurements are difficult to interpret because growth at several sites is included in a single dimension. With the mandible as an example, Garn noted that various parts of this bone must be considered separately because their growth is only moderately intercorrelated. Garn commented further on the independence of tooth formation from general somatic development and caloric intake.

Few studies have been made at Fels of the growth of the facial skeleton, other than the mandible, but Ohtsuki *et al.* (1982a) reported that the angle S–N–A, which is an indicator of maxillary position relative to the cranial base, decreased significantly in each sex until about 10 years; later it tended to increase. Low values for this angle indicate a more posterior position of the central part of the face.

There have been, however, several studies of the growth and development of the mandible. This concentration on the mandible partly reflects the presence of better landmarks for this bone than for the remainder of the face. In agreement with the views of Garn (1961a[57]), these studies of the mandible were related to specific regions of the bone. Israel (1969, 1978) analyzed the height of the body, cortical thickness near the premolar teeth and the distance from the apex of the second premolar to the superior and inferior margins of the body. His cross-sectional data for ages from 6 to 25 years showed that body height, in which there was a pubescent spurt, and cortical thickness were similar in the two sexes to 10 years but they were significantly larger in boys at older ages (Fig. 6.8). Men exceeded women by about 20% in cortical thickness and 13% in body height. This latter difference was associated with a larger distance from the apex of the premolar tooth to the inferior margin of the mandible, but the distance from the apex of this tooth to the superior margin of the mandible did not differ between the sexes.

A study of the height and anteroposterior depth of the mandibular symphysis was reported by Garn, Lewis and Vicinus (1963d[127]) who related their findings to the mesiodistal diameters of the central incisor and first molar teeth, and to adult values for stature and bony chest breadth. The symphyseal dimensions were effectively independent of the other recorded variables in adulthood except for low correlations between the height and depth of the symphysis and between both these measurements and tooth size. These workers separated adults into groups depending on whether symphyseal height or symphyseal depth was greater or less than the median. There were no significant tendencies to assortative mating for

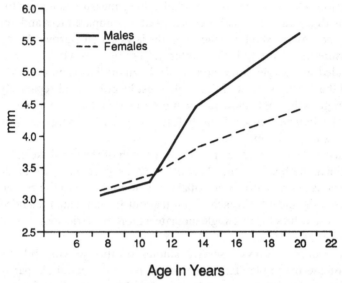

Fig. 6.8 Means for the cortical thickness of the mandible at the inferior margin of the body near the premolar teeth. (Israel, H. (1969). Pubertal influence upon the growth and sexual differentiation of the human mandible. *Archives of Oral Biology*, **14**, 583–90. Redrawn with permission from Pergamon Press, plc, copyright 1969.)

symphyseal size. The offspring of low × low and high × high matings differed in symphyseal height by about 2 mm which corresponded to about 40 percentile points. There was a similar difference in symphyseal depth between the offspring of thick × thick and thin × thin matings. Further-more, the offspring of thick × thin and thin × thin matings were similar in symphyseal depth which suggested a form of inheritance in which thin is dominant over thick.

Heiber (1975) related the growth of the mandible in boys to the onset of ossification in the ulnar sesamoid and pubescent spurts in stature. He found a statistically significant increase in the rate of growth in S–Gn, which is a measure of chin position relative to the cranial base, that tended to parallel the pubescent spurt in stature. The overall length of the mandible (Ar–Gn) and Ba–Gn had pubescent spurts about 1 year after the onset of ossification in the ulnar sesamoid. These spurts tended to be later than those in S–Gn.

The growth of the mandible during pubescence was described in more detail by Lewis and co-workers (1982). In their data, there were spurts in three major dimensions of the mandible (Ar–Go, ramus height; Go–Gn, body length; and Ar–Gn overall length) in most children but they were less common in girls than boys and less common in ramus height than in the other dimensions. The spurts tended to occur about 1.5 to 2.0 years later in boys than in girls and were about 33% larger in boys. The mean increments in mandibular dimensions decreased before the spurts occurred. The spurts tended to occur several months before peak height velocity and about 1 year after ossification of the ulnar sesamoid, but before menarche. Girls with early menarche tended to have larger mandibular spurts than those with late menarche, but the relationships between the prevalence of spurts and the rate of maturation were not consistent. The timing of spurts was more variable relative to ulnar sesamoid ossification than in relation to peak height velocity.

Boys with large spurts in all three mandibular dimensions tended to be late in ulnar sesamoid ossification but there was an opposite tendency in girls. Boys who passed rapidly through pubescence had small spurts in ramus length and body length but not in overall length, but girls who passed rapidly through pubescence tended to have larger spurts in all three dimensions. There were low correlations in the size of the spurts in the three dimensions considered.

Changes with age in the angle of the mandible (Ar–Go–Gn) in women aged 26 to 90 years were analyzed by Israel (1973a, 1973c). Each participant had a radiograph when aged at least 25 years and another after an interval of 14 years or longer. The difference in the mandibular angle between those who were edentulous and those with enough teeth to allow occlusion was not significant. Serial data did not demonstrate changes of more than 2.5°, nor did they reveal the marked changes in mandibular shape with aging that are commonly illustrated in textbooks.

### Third cervical vertebra
Israel (1973d) using cross-sectional and serial data described increases with age in the width but not the height of the body of the third cervical vertebra in women aged 26–90 years. Israel considered this was part of a widespread remodelling of the skeleton during adulthood that includes increases in the breadth of the shafts of long bones and redistribution of bone in the craniofacial area.

### Pelvis
Several major studies of the growth of the pelvis have been made at Fels. Indeed, one could claim that these are the only comprehensive studies of pelvic growth that have been reported. Reynolds (1945, 1946b) described pelvic growth from early infancy to 9 years. Many of the distances and angles measured may be unfamiliar. Those mentioned in this text are

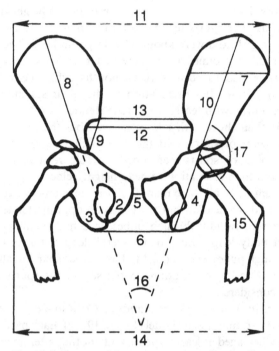

Fig. 6.9 A diagram of the measurements made on radiographs of the pelvis by Reynolds (1945, 1947): 1 = superior pubic ramus length, 2 = inferior pubic ramus length, 3 = inferior ischial ramus length, 4 = height of body of ischium, 5 = interpubic breadth, 6 = interischial breadth, 7 = ilial breadth, 8 = iliac length, 9 = breadth of greater sciatic notch, 10 = pelvic height, 11 = pelvic breadth, 12 = pelvic inlet breadth, 13 = interiliac breadth, 14 = bitrochanteric breadth, 15 = length of femoral neck, 16 = pelvic angle, and 17 = femoro–pelvic angle.

illustrated in Fig. 6.9. Reynolds reported that radiographic and anthropometric data showed matching age trends, although the latter were consistently larger. At birth, the superior pubic ramus, the inferior ischial ramus and much of the body of the ischium were ossified and there were some centers of ossification in the sacrum.

Reynolds divided his measurements into those that are parallel to the radiographic film and, therefore, only slightly distorted, and radiographic abstractions such as 'ischial length.' Various pelvic dimensions were highly intercorrelated at birth except for interpubic breadth, interischial breadth and iliac breadth. Later, during infancy, the intercorrelations between measurements were generally about 0.4 and tended to be higher in boys than in girls (Reynolds, 1945).

There were wide ranges of values at birth and rapid growth to 3 months that was succeeded by progressive deceleration. There were only small sex differences in pelvic measurements during infancy, but there were some changes in shape during infancy. The iliac index (breadth/length) increased markedly to 1 month and then decreased slowly to 9 months, after which it increased until it reached about the same value as at birth. The greater sciatic notch widened to 1 month and narrowed from then to 3 months, after which it widened slowly to 1 year.

There was considerable tracking of growth during the first year: values at birth and 1 year were correlated about 0.6, except for the iliac breadth and the breadth of the greater sciatic notch for which the correlations were near zero. The iliac index and the height and breadth of the pelvis tended to be greater in boys than girls, but girls were larger in inlet breadth, interpubic breadth, interischial breadth, the greater sciatic notch, and pubic length. This led to the conclusion that boys tended to have large outer pelvic structures, while girls tended to have larger inner pelvic structures.

At birth, pelvic breadth was correlated with body weight and recumbent length in each sex and with weight/recumbent length$^3$ in girls but not boys. The inlet index of the pelvis (breadth/depth) and the cephalic index (cranial length/cranial breadth) were not significantly correlated, but inlet breadth and head circumference at birth were correlated (r = 0.5). In boys, but not girls, pelvic height was correlated with a skeletal maturity index derived from the number of ossified centers and with the age at emergence of the first tooth, but not with age at first walking. There were significant correlations between mothers and infants for the breadth, but not the depth, of the pelvis. Despite this, the correlations between siblings for pelvic size were not significantly larger than those between non-siblings.

Reynolds (1946b) described the growth of the pelvis from 1 to 9 years. He found few significant lateral differences, and based his report on the means for the two sides. An intercorrelation matrix at 34 months showed very close relationships within a group of measurements that included many breadths, ischial and pubic lengths, and the length of the neck of the femur. Another group of significantly interrelated variables included the breadth and depth of the pelvic inlet, the breadth and height of the pelvis, and iliac length. Unfortunately, these groupings were based on subjective judgments without the benefit of factor analysis, although this had been employed earlier at Fels by Robinow (1942b). Additionally, the breadth of the greater sciatic notch was related to the depth of the pelvic inlet and iliac length. In general, the intercorrelations at 34 months were slightly higher than those during early infancy and they tended to be larger in girls than in boys (Reynolds, 1945).

Generally, the growth rate decelerated from 1 to 9 years for most pelvic dimensions, but the pelvic angle was almost constant during this age interval. Boys tended to have significantly larger values than girls for pelvis height and breadth, iliac length, interiliac breadth, bitrochanteric breadth, length of the femoral neck and pelvic angle. The girls tended to have higher values for the inlet index, interpubic and interischial breadths, pubic length, breadth of the sciatic notch and the femoro-pelvic angle. As in infancy, boys had larger outer pelvic structures and girls had larger inner structures, but the sex differences tended to decrease with age. Reynolds (1946b) reported that those with large pelvises from 3 to 9 years tended to have greater trunk and limb circumferences and more subcutaneous adipose tissues. Also, there was evidence of relationships between pelvic size and maturity; those with large pelvises tended to be advanced in ossification of the proximal femoral epiphysis and the completion of the obturator foramen.

The age range of Reynolds' studies was extended by Coleman (1969), who analyzed data from 9 to 18 years. His main interest was in the apparent mechanisms by which the pelvis reaches its adult form, which involves marked sex differences. On tracings of radiographs, he drew vertical and horizontal reference planes aligned on the maximum bi-iliac breadth and the midpoint of the pubic symphysis. He used X and Y coordinates to locate landmarks that reproduced the pelvic outline, and fitted orthogonal polynomials to the serial coordinates after serial outlines for individuals were superimposed at their common mean points and rotated to minimize the differences within sets of serial data.

Distance curves were plotted for each point as a function of time. Almost all of the point locations had pubescent growth spurts. Coleman showed that the sex differences in the growth of the pelvis were due to variations in rates and directions of growth at local areas, particularly the greater growth of the internal acetabular and pubic regions in females. The middle part of the iliac crest tended to curve less laterally in females than in males at young ages. During growth, the iliac crest moved superiorly and laterally in each sex, but during the growth of this and the acetabulum, the lateral direction predominated in females. There was general remodelling and relocation over the ischial tuberosity in each sex, but this tended to be more laterally directed in females. There was more growth at the superior and medial borders of the pubis in females. With growth, the point at the greatest breadth of the pelvic inlet moved laterally and inferiorly in females, but in males it moved laterally and superiorly (Coleman, 1969).

This study by Coleman contributed to knowledge of sexual differentiation of the pelvis, particularly the pelvic inlet. For example, the inlet

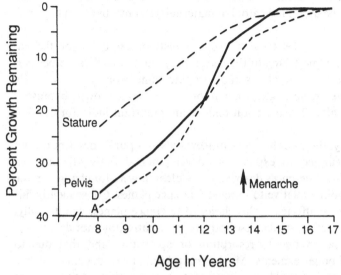

Fig. 6.10 The percentage of growth remaining for stature and pelvic breadths in girls at ages 8 to 17 years (A = inlet breadth; D = interischial breadth). Data from Moerman (1982).

breadth was about 1 cm greater in females than males at 18 years, due to more lateral growth of the acetabulum and pubis. The greatest breadth of the female inlet moved inferiorly because of the lateral growth of the ischia, but in males the upper half was the broader part.

Another study of the growth of the pelvis in females aged 8 to 18 years was made by Moerman (1981, 1982). She found that pelvic breadths increased more slowly than stature, but the growth of the pelvis continued to at least 18 years (Fig. 6.10). Much smaller percentages of size at 18 years were reached by pelvic breadths than by stature at ages 8 to 10 years, but the difference was reduced during the next few years due to large pubescent spurts in the pelvis.

In the Fels data, pelvic breadths 3 years after menarche did not differ significantly from those at 18 years (Moerman, 1982). Rapidly maturing girls, compared with slowly maturing girls of the same gynecological age, were smaller in general body dimensions and in pelvic size at menarche, but there were only minor differences between these groups at 18 years. From these findings, Moerman concluded that some risks of teenage pregnancy may be associated with early menarche and incomplete pelvic growth. In a later analysis of her data, Moerman (Lavelle, 1991) found decreases with age in the correlations between the width of the pelvic birth canal and either stature or bicristal diameter. She concluded that hip width and stature might be useful indicators of maternal–fetal morbidity in high-risk populations.

Chumlea (1983[3], 1984[4]) reviewed the histological basis for the enlargement and remodelling of the pelvis. He pointed out that the pelvis is three-dimensional, but all its radiographic dimensions are linear. He noted that the pelvic angle is not informative about growth because it changes little after 2 years, but it may be an important indicator of adult shape.

In summary, the valuable work of Reynolds was purely descriptive, but Coleman attempted to explain the developmental basis of adult sex differences. The differences described by Coleman were for changes in the locations of points relative to selected reference planes. Consequently, his conclusions were affected by his choice of reference planes. Additionally, Coleman's procedure, while descriptively informative, did not distinguish between relocation due to resorption or apposition and that due to movement of bone segments. Moerman, in her studies, documented late growth changes that could have important obstetric effects. At the time of her work, its obstetrical applications were recognized, but its application to the design of automobile restraint devices was not recognized until much later (Chumlea, 1984[4]). This is a most important topic area because automobile injuries are a major cause of death during childhood.

### Skeletal variations

A review of skeletal variations based largely on Fels data has been published by Garn (1966a[66]) who dealt with their genetic control. He pointed out that genetic variability may be maintained by differing directions of selection within a population. One genotype may be adaptive in childhood and another may be preferable at older ages, and in addition, some mechanisms may favor heterozygotes. Garn considered that descriptions of variations should be followed by attempts to apportion their causation to groups of genetic and environmental influences. In other reviews of skeletal variations, Roche (1978c[186], 1986b[202]) emphasized their importance in relation to the clinical applications of reference data.

#### *Variations in ossification*

In an early Fels study, attention was directed to a common variation in the pattern of calcification of the distal femoral epiphysis (Sontag & Pyle, 1941a[294]). In many children, the margin of this epiphysis was irregular, especially on its medial aspect, and calcified areas extended beyond this part of the margin without changes from normal in the density of the epiphysis. This condition, which appears to be associated with a broad calcified zone, is known as disseminated calcification. It was very common at 1 to 3 years when the distal femoral epiphysis was enlarging rapidly, but it is not useful as a maturity indicator because it does not occur in all children. It usually appeared earlier in girls than in boys and in those who matured rapidly or were young when they first walked. This condition was of shorter duration in those who were growing rapidly. The condition generally lasted 1 to 6 years, but most of the irregularity was present for less than 2 years. Corresponding less severe changes occurred in the proximal epiphysis of the tibia and in the tarsal bones.

Missing secondary ossification centers (epiphyses) were shown to be common in the foot (Garn et al., 1965e[153]). Individuals varied widely in this respect, but there was considerable sibling concordance that was partly explicable in terms of X-linked inheritance. Garn pointed out that when centers were missing, the nearby bones tended to fuse, with a consequent decrease in flexibility and, perhaps, in fine movements.

When pseudoepiphyses were present, the ossified epiphysis was joined to the shaft by bone long before the usual age for epiphyseal fusion. Consequently, elongation at the epiphyseal zone stopped at a very early age. Pseudoepiphyses were relatively common at the non-epiphyseal ends of the short bones of the hand and foot and they tended to be concentrated within families (Garn & Rohmann, 1966b[142]).

Notches near the non-epiphyseal ends of the metacarpals are a mild form of pseudoepiphyses (Lee & Garn, 1967). These notches have special interest because of their potential for use as natural bone markers from which elongation can be measured (Lee, 1967). In a survey of Fels participants, Lee, Garn and Rohmann (1968) found that notches occurred at multiple sites in normal children. They were common at the radial side of the base of the second metacarpal (30%) and the ulnar side of the base of the fifth metacarpal (88%). Each notch was graded as slight, moderate or marked, and a score was assigned to each metacarpal that was the total for its radial and ulnar sides. Because the scores tended to change with age, the maximum for each individual was used in subsequent analyses that showed notching of the second metacarpal tended to be associated with stature, but in ways that differed between the sexes. In males, there was a negative correlation that was significant at 3 years only, while in females there was a positive association that was significant in adults only. The degree of notching was not related to the rate of maturation, except that males with marked notching of the second metacarpal tended to have earlier fusion of the distal phalanges.

In a minority of children, some epiphyses of the bones of the hand and foot were conical, with the apex of the cone directed towards the end of the shaft where there was a corresponding indentation. Conical epiphyses were more common in females than males (Hertzog, Garn & Church, 1968). In the hand, they were most common on the fifth middle phalanx and the first distal phalanx. They occurred at both these sites in some children; this combination was more common than would be expected due to chance, especially in females. It was shown by Hertzog that the shaft of the fifth middle phalanx is absolutely and relatively short prior to the development of a conical epiphysis. The ages at ossification and of fusion of conical epiphyses varied considerably, but fusion did not tend to occur early. This marked variability in timing indicates that this trait could affect the lengths of bones in which it occurs.

### Brachymesophalangia
It has been noted that the fifth middle phalanx of the hand is commonly short and broad (brachymesophalangia). This condition is often associated with radial angulation of the fifth finger (clinodactyly). Brachymesophalangia was present in 0.6% of the Fels participants, in whom it appeared to be inherited in a dominant fashion (Garn, Fels & Israel, 1967h[79]; Hertzog, 1967; Garn et al., 1972[93]; Poznanski et al., 1972).

### Variations of joints
Selby and colleagues (1955) described the prevalence and the familial nature of a bony bridge on the first cervical vertebra (atlas). This bridge crossed the groove for the first cervical nerve, the vertebral artery and its accompanying veins on the superior surface of the posterior arch of the atlas. There was a complete or incomplete bridge in 27% of the Fels participants who were aged more than 14 years; about half of these bridges were complete. Sex differences in prevalence were small. The mean age of appearance was 9 years. Familial tendencies were evident, although pedigree analysis showed that bridging may not be expressed in the next generation, and it may occur in children although it was absent in their parents. Selby considered this bony bridge might be inherited in a Mendelian dominant fashion.

Poznanski et al. (1985) reported reference data for some measurements of the knee joint spaces because these are reduced in juvenile arthritis. The apparent joint space was best measured in the midline in children up to 5 years, but on the medial side in older children. When used in combination with joint widths, these measurements assisted the detection of early stages of juvenile arthritis.

Fusion of the fifth lumbar vertebra to the first sacral vertebra was present in about 16% of the Fels participants (Garn & Shamir, 1958[164]). It showed high concordance within sibling pairs and parent–offspring pairs.

### Cysts

Benign asymptomatic cyst-like areas near the distal end of the shaft of the femur were described by Sontag and Pyle (1941b[295]). These areas rarely extended into the epiphyseal zone but maintained a constant distance from it. They were usually round or oval and some were sacculated. They were up to 3 cm in diameter with sharply defined borders, and were positioned just beneath the cortex posteriorly. The average age of appearance was 46 months and they soon reached their maximum size. On average, they were present for 29 months, being more common in boys (53%) than in girls (22%). Sontag and Pyle did not find evidence of trauma. They speculated that these cysts might be due to the presence of small cartilaginous nests in rapidly growing areas.

These cysts were investigated later by Selby (1961). He reported that the first sign of their development was a thickening of trabeculae, but sometimes they appeared at a site separated from the area of thickening. In Selby's data, the cysts remained for 1 to 11 years, which was much longer than the duration reported by Sontag and Pyle (1941b[295]). Additionally, while the prevalence in girls (29%) was slightly greater, the prevalence in boys (27%) was much less than that reported by Sontag and Pyle. Selby noted some evidence of a familial tendency.

### Growth arrest lines

Growth arrest lines occur near the epiphyseal zones of long and short bones in some children. They reflect alterations in the architecture of the cancellous bone that may be associated with disease and slowing of growth. In an interesting paper, Sontag (1938[276]) was the first to describe opaque bands near the centers of ossification of some tarsal bones. These bands were similar radiographically to growth arrest lines. They were 1–2 mm wide and occurred about where the margin would have been in the neonatal period. Sontag considered that these bands may reflect a slowing of growth soon after birth. They were not due to infections, but they were common in the infants of malnourished mothers (Sontag, 1938[276]; Sontag & Harris, 1938[287]).

Growth arrest lines near the distal end of the tibia are very common in Fels participants (Garn, Hempy & Schwager, 1968b[92]). Since these lines may persist into the ninth decade, particularly in females, they form natural bone markers from which the contributions of the proximal and distal epiphyseal zones to the total length of the bone can be determined

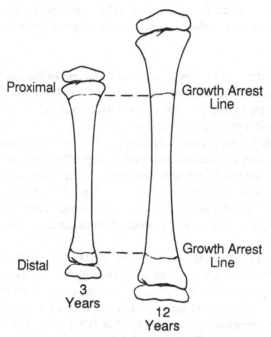

Fig. 6.11 A diagram to illustrate the use of growth arrest lines near the proximal and distal ends of the tibia at 3 and 12 years in the measurement of elongation at the proximal and distal epiphyseal zones. Since the lines are fixed, the distances from them to the ends of the bone increase due to elongation at the epiphyseal zones.

(Garn & Schwager, 1967[160]; Gindhart, 1969). Serial radiographs showed that some lines were resorbed, but others formed before this resorption occurred. Consequently, long series of measurements could be constructed from groups of lines overlapping in duration.

This method of using growth arrest lines is shown in Fig. 6.11 where the outlines of the tibial shaft for an individual at 3 and at 12 years are aligned on growth arrest lines near each end of the shaft. The positions of the proximal and distal growth arrest lines at 3 years can be projected onto the outline of the shaft at 12 years. As a result, measurements can be made of the elongation during the interval from 3 to 12 years at the proximal end and, separately, at the distal end. Garn and his co-workers concluded that slightly less than half the total elongation of the tibia occurred at its distal end and that this proportion increased slightly from 2 to 10 years. Additionally, growth arrest lines can be used to show the relative rates of apposition and resorption on the medial and lateral aspects of the shaft of the tibia.

Gindhart (1969) reported that almost all the Fels participants developed growth arrest lines at the distal end of the tibia before 13 years, but none appeared after this age. The development of new lines was most common from 1.5 to 3.0 years. There were low, but significant, associations between the formation of lines and the occurrence of infectious diseases, but not with tonsillectomies. The correlations between the number of arrest lines and adult stature were not significant. Arrest lines in the tibia were present in 14% of men and 30% of women aged 25 to 50 years and in 8% of men and 14% of women aged 51 to 86 years (Garn & Schwager, 1967[160]). The decrease with age was due to cortical remodelling at the endosteal and periosteal surfaces. Consequently, those in whom the lines were resorbed tended to have wider marrow cavities.

Growth arrest lines in a set of monozygous triplets were described by Sontag and Comstock (1938[284]). They estimated the ages at which the lines formed from the distances between the lines and the ends of the tibia, fibula, femur and radius. They could not identify a single factor that was associated with all the lines and they concluded that the differences between these monozygous triplets in their growth arrest lines demonstrated the involvement of environmental mechanisms.

### Skeletal mass

A large amount of work at Fels has been related to skeletal mass. The early work was based on measurements of the cortical thickness of the second metacarpal. This bone was chosen because many hand–wrist radiographs were available, and the central part of the shaft is nearly circular in cross-section. The latter quality simplifies the derivation of areas from measurements of cortical thickness (Garn, 1963a[63], 1963b[64]). The cortical thickness of this bone is only an approximate index of skeletal mass despite the positive correlations between these values. Realization of the limitations of cortical thickness measurements led to the development of radiographic densitometry. In this technique, the amount of bone mineral in a transverse slice of a bone is estimated from light transmitted through a radiograph. This procedure was used extensively at Fels and numerous technical improvements were developed.

Both these methods provided local indices of skeletal mass, which is important in relation to osteoporosis. Somewhat surprisingly, Garn (1963a[63]) concluded that 'accurate estimates of bone weight are superfluous in obesity studies, where the loss of obesity tissue may exceed the total weight of the skeleton.' This is true for studies of changes in body weight, but for studies of fat-free mass or total body fat calculated from body density the influence of changes in skeletal mass is critical.

### Cortical thickness

Cortical thicknesses were measured at Fels using fine-pointed dial calipers that were read to the nearest 0.1 mm. All these measurements were highly replicable, although replicability was higher for the total breadth of the bone (periosteal breadth) than for the breadth of the marrow cavity (Garn, 1970[70]). With the introduction of electronic digitizers, it became important to compare replicability between measurements made with this new equipment and those made with calipers. Both sets of data showed high replicability, but the intra-observer errors with a digitizer were much smaller than those with calipers. The systematic differences between these sets of measurements were very small (Chumlea, Mukherjee & Roche, 1984a[25]).

The advantages and limitations of cortical thickness measurements have been considered in several reviews. Few procedures other than the measurement of cortical thickness can provide information about long term serial changes (Roche, 1987b[205]). The changes in cortical thickness and, by inference, in skeletal mass, during childhood and pubescence are important because they are major determinants of skeletal mass in young adulthood. In turn, values in young adulthood play a major role in determining values for skeletal mass in middle and old age.

Roche (1978c[186], 1986b[202]) provided sources of reference data and described histological mechanisms that affected the total bone breadth and the breadth of the marrow cavity and, consequently, cortical thickness. Apposition on the external (subperiosteal) surface of a bone is due to the actions of osteoblasts in the deep layer of the periosteum. At the same time, bone is resorbed from the endosteal surface and, therefore, the marrow cavity widens. The balance between these changes determines the cortical thickness which may increase more in some parts of a bone than in others. Additionally, the cortex may drift medially or laterally due to remodelling. This can be demonstrated histologically or by changes in growth arrest lines (Garn, Davila & Rohmann, 1968f[78]). Reference data for the cortical thickness of the second metacarpal, and values derived from it, have been published by Garn et al. (1970[156]). Falkner (1975) stressed the need to take ethnic differences into account when applying these reference data.

In early Fels work, it was shown that the total bone width, relative to the total width of the calf, was greater in males than females at all ages from birth to young adulthood (Reynolds, 1948; Reynolds & Grote, 1948). In the tibia, cortical thickness tended to be greater in boys than in girls except during pubescence when some reversals occurred due to marked endosteal apposition in girls. The cortical thicknesses of the tibia were correlated with those of the humerus, but neither of these measurements necessarily reflects skeletal mass. With aging, cortical thickness seemed to decrease proportionately less in the tibia than in the second metacarpal (Garn, Rohmann & Nolan, 1963e[150]; Garn *et al.*, 1964c[151], 1967g[155]; Spencer, Sagel & Garn, 1968). Cortical thickness of the second metacarpal increased during childhood because subperiosteal apposition exceeded endosteal resorption (Garn et al., 1963e[150]; Frisancho, Garn & Ascoli, 1970; Garn & Wagner, 1969[170]; Fig. 6.12). Endosteal resorption, as shown by increases in the breadth of the marrow cavity, continued until about 12 years in girls and 16 years in boys. Later, there was marked endosteal apposition in girls which reduced the breadth of the marrow cavity of the second metacarpal (Garn *et al.*, 1968h[171]; Garn, 1970[70]).

As a result of these changes, cortical thickness of the second metacarpal increased to about 25 years and then remained constant until about 45 years (Garn *et al.*, 1969b[157], 1970[156]). In both sexes decreases occurred after this age, so that by the late sixties the mean values were less than those during adolescence (Fig. 6.13). During adulthood, cortical thickness of the second metacarpal tended to be greater in men than women at all ages and it decreased more slowly in men than in women after 45 years. This decrease was due to resorption at the endosteal surface that began at about 40 years. It continued at a steady rate in women but the rate of loss decreased in men (Spencer & Coulombe, 1966; Spencer, Garn & Coulombe, 1966; Garn *et al.*, 1964c[151], 1967g[155]). Despite these common changes, the loss was small in some individuals, many of them tall, because increased periosteal apposition compensated for endosteal resorption.

From comparisons between Fels participants and poorly nourished children and adults from Central America, and from the analyses of pedigree data available at Fels, it was concluded that cortical thickness was determined more by genetic influences than dietary ones (Garn & Rohmann, 1964a[139], 1966a[141]; Garn, 1961c[59], 1966c[68], 1970[70]; Wolánski, 1966b; Falkner, 1975). For example, the correlations between cortical thickness and calcium intake were not significant (Garn, 1962a[60], 1967[69]; Garn, Pao & Rohmann, 1965d[132]). The low, but significant correlations for total bone width and cortical thickness within parent–offspring pairs and sibling pairs have a pattern consistent with a hypothesis of X-linked inheritance (Garn *et al.*, 1967g[155]).

Spencer *et al.* (1966) fitted a model to the mean cortical thicknesses at each age. This model, that fitted well, postulated that the value at an age was a balance between apposition and resorption.

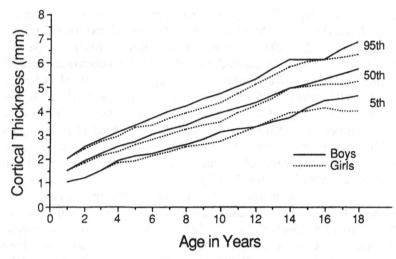

Fig. 6.12 Selected percentiles for the cortical thickness (mm) of the second metacarpal (data from Garn *et al.*, 1967e[154]).

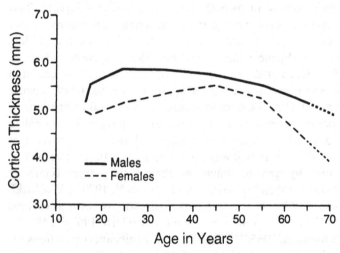

Fig. 6.13 Mean cortical thickness of the second metacarpal from 12 to 70 years. (Garn, S.M., Rohmann, C.G. & Nolan, P., Jr. (1964c). The developmental nature of bone changes during aging. In *Relations of Development and Aging*, ed. J.E. Birren, pp. 44–61[151]. Redrawn, courtesy of Charles C Thomas, Springfield, Illinois.)

### Cortical area

If it is assumed that the cross-section is circular, the total breadth and the cortical thickness of the second metacarpal can be used to calculate cortical area. In Fels data, these areas were larger in boys than girls at all ages and they showed spurts, from 12 to 14 years in boys and 10 to 12 in girls, that were considerably larger in boys than girls. About 23% of the increase in cortical area during pubescence was due to endosteal apposition in males compared with 36% in females. Cortical area, expressed as a percentage of the total cross-sectional area, was similar in the two sexes until 12 years, after which it was slightly larger in females than males (Frisancho *et al.*, 1970). The cross-sectional cortical area of the second metacarpal increased rapidly until about 20 years in males and 11 years in females. From 40 to 90 years, the total loss in this area was about 15% in men and 39% in women (Garn *et al.*, 1964c[151]). This loss was due to endosteal resorption; there was an increase in the total cross-sectional area of the second metacarpal (Rohmann *et al.*, 1967; Garn *et al.*, 1968h[171]). The proportion of the total cross-sectional area occupied by cortical bone was greater in women than in men, at least to 60 years. These findings from cross-sectional data were supported from data obtained after a 15-year interval during adulthood. These follow-up data also showed the loss of cortical thickness was correlated with the initial value (0.4).

Garn (1970[70]) reported an index of skeletal mass derived from the cross-sectional cortical area of the second metacarpal or tibia, and the length of the corresponding bone. By relating this index in adults to known population values for skeletal mass, he derived a proportionality constant (coefficient) to allow direct estimates of skeletal mass. This method has not been applied widely because of justifiable concerns about the limited representativeness of either the second metacarpal or the tibia. Garn, however, applied this method and published reference estimates. In turn, these estimates were used to calculate differences between annual means of skeletal mass from which Garn derived increases in the body stores of bone mineral (mg/day). This led to estimates of calcium retention/day that were about 100 mg/day, except during adolescence when there was a retention of about 300 mg/day by boys and about 200 mg/day by girls. Applying this approach, Garn and Wagner (1969[170]) claimed that skeletal mass increased by about 2% after 18 years. While much of this work remains interesting and relevant, better methods are now available for the measurement of skeletal mass.

### Medullary cavity width

Garn *et al.* (1968f[78]) directed attention to 'normal' medullary stenosis in which the breadth of the marrow cavity of the second metacarpal is less than 1.0 mm. This condition occurs in some healthy individuals in whom it is probably inherited as a simple Mendelian dominant trait. At Fels, the men with the narrowest marrow cavities had low values at birth that decreased through childhood and early adulthood due to endosteal apposition. The women with the narrowest marrow cavities had marked endosteal apposition during adolescence. In those with medullary stenosis, the second metacarpal was unusually narrow, but cortical thickness values were high.

Rapidly maturing boys and girls had wider bones but the differences from children maturing at average rates were more marked in girls than boys (Reynolds, 1946a, 1948; Reynolds & Grote, 1948). In young adulthood, tall individuals tended to have slightly more periosteal apposition, less endosteal resorption and increased cortical thickness. At ages older than 45 years, the total breadth of the second metacarpal tended to be greater and the decrease in breadth tended to be smaller in those who were tall (Garn & Hull, 1966[94]).

### Radiographic densitometry

Radiographic densitometry is the measurement of the ability of a radiograph to prevent the transmission of light. The parts of a radiograph that transmit large amounts of light are those that were shielded from x-rays by dense objects, especially bones. The major application of radiographic densitometry concerns the estimation of skeletal mass. Garn *et al.* (1966c[80]) compared cortical thickness measurements and densitometry for the measurement of bone loss. Cortical thickness measurements were more suitable for tubular bones than for those that were mainly cancellous. Variations in radiographic techniques were much less critical for cortical thickness than for densitometry for which non-screen film and highly standardized processing are mandatory. There is also some 'hardening' of the x-ray beam as it passes through tissues. Garn and his associates considered that caliper measurements slightly overestimated the total breadth and underestimated the breadth of the marrow cavity. Colbert *et al.* (1980) reported correlations of about 0.6 between cortical area obtained using a densitometer, estimates of the total bone mineral of the second metacarpal, and bone mineral expressed per centimeter of the length of this bone (bone mineral content). It was shown that measurements of cortical area were insensitive to bone loss due to the poor precision of the measurements.

Densitometry can be used to compute bone mineral content if the cross-sectional area of the bone is known; subsequently the ratio of bone mineral to the volume of bone tissue can be used as an index of mineralization (Colbert, 1968, 1974; Colbert & Garrett, 1969). This mineralization index can help monitor change in an individual, but not to calculate absolute bone mineral concentration or to compare individuals. A simple theoretical model for determining the absorption coefficient of bone specimens of regular or irregular shape *in vitro* was proposed by Colbert, Spruit and Davila (1967). This allowed the derivation of important parameters that were independent of the wave length of the x-rays. The results obtained with this model were in substantial agreement with observed data. Bone mineral content from densitometry was highly correlated (0.9) with ash weight and elemental calcium; the latter can be predicted with errors of about 6%. Data from densitometry were also highly correlated with those from single photon absorptiometry (Colbert, Schmidt & Mazess, 1969; Colbert, Spruit & Davila, 1970a; Colbert, Mazess & Schmidt, 1970b).

Much of the work done at Fels by the Colbert group was performed on the fifth middle phalanx. This is a convenient bone to scan, but it may have been a poor choice for biological reasons. Since abnormalities and variations of this bone are common, it is unlikely to be representative of the skeleton (Garn *et al.*, 1967h[79]). Colbert, Van Hulst and Spruit (1970c) provided graphs of reference values for bone areas measured directly on radiographs, estimated bone weights and estimated bone densities of the first distal phalanx and middle phalanges II–V of the hand.

Colbert (1968) applied his approach to bones of irregular shape. He showed that the measured areas were not closely related to the estimated bone weights until after 5 years and that the areas were not closely related to density until after 20 years. The measured bone sizes increased rapidly to 15 years and then slowly, whereas the estimated bone weights increased rapidly to 17 years. Bone density increased rapidly to 13 years; this was followed by a pubescent spurt to 17 years and a continuing slow increase to 35 years.

Israel (1968b) demonstrated that densitometry could provide an index of skeletal mass in a precise area, despite problems due to non-linear relationships between radiographic density and radiographic exposure, and variations in film type and processing. Other errors are produced by the polychromatic nature of the radiographic beam and by its scattering. Despite these limitations, a densitometric trace across a bone demonstrates its fine structure (Israel, 1967b; Israel, Garn & Colbert, 1967). Using this technique, Israel documented a 14-year persistence of trabeculae near the roots of incisor teeth and used these to document the loss of alveolar bone. He used the same approach to study changes in the calcaneum where trace patterns could be matched over time, if one allowed for stretching and compression of the image due to variations in positioning. The possible application of densitometry to dental enamel and dentine and to studies of craniofacial remodelling was noted (Colbert, Israel & Garn, 1966; Israel, 1967b).

### Plasma alkaline phosphatase

Plasma alkaline phosphatase was studied by Clark and his associates (Clark & Beck, 1950; Clark, Beck & Shock, 1951a; Clark, Beck & Thompson, 1951b) because of its known relationship to osteoblastic activity and, therefore, the rate of bone growth and remodelling. In children and young adults, there was a marked consistency of measurements from year–to–year (r = 0.6–0.8), but marked variation in these measurements from hour to hour and from day to day. The plasma alkaline phosphatase values were high in infancy, but decreased near the end of the first year. Sex differences and age changes were slight until marked abrupt decreases occurred at 12 years in girls and 14 years in boys. The levels of this enzyme tended to parallel the rate of growth in stature. During pregnancy, women with male fetuses tended to have higher values than those with female fetuses, and there was a two- or three-fold increase during the last trimester of pregnancy that was not affected by lactation. In adults, the values were lower in males than females but they increased in old age, particularly in males, perhaps due to prostatic enlargement (Clark *et al.*, 1951b).

### Teeth

An enormous body of work on the size and shape of the permanent teeth was performed by Fels scientists led by Garn and Lewis. Their careful measurements of dental models provided important descriptive information and many insights into relationships within the dentition and between the dentition and the remainder of the body. As a result of this work, our understanding of the determinants of tooth size and shape is greatly enhanced.

#### Tooth size

Reference data, including maximum confidence values, for the mesiodistal dimensions of the permanent teeth were published (Garn, Lewis & Walenga, 1968a[128]). Some of the these measurements were made using a transducer caliper (Garn, Helmrich & Lewis, 1967i[91]). In these data, lateral differences were small and inconsistent, both within individuals and on a group basis. The sex difference in mesiodistal diameter was about 4% for most teeth, but was larger for the canines and the teeth near them. Garn and his colleagues considered these variations in the sex differences constituted evidence of a canine 'field' for size difference between the sexes (Garn, Swindler & Kerewsky, 1966d[169]). The largest absolute sex difference was for the mandibular first and second molars followed by the canines; it was smallest for the incisors. The sex difference in tooth size was larger,

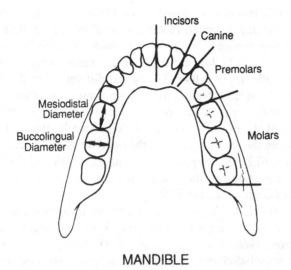

MANDIBLE

Fig. 6.14 A diagram to show the occlusal aspect of the groups of mandibular teeth and examples of their mesiodistal and buccolingual diameters.

about 6%, for the buccolingual diameter than for the mesiodistal diameter (Garn, Lewis & Kerewsky, 1964b[106], 1967b[112]). These diameters are illustrated in Fig. 6.14.

Garn *et al.* (1965c[118]) and Garn, Lewis and Kerewsky (1965f[109]) reported that communality indices for the mesiodistal diameter tended to be lower for the distal teeth in each morphological class than for the other teeth within these classes. The distal teeth also had lower size interrelationships with each other than did the more mesial teeth. This finding was not due to greater size variability of the more distal teeth. The lack of negative associations between adjacent teeth in mesiodistal diameters indicated that size variation was not dependent upon variations between teeth in their shares of the anlage (Garn *et al.*, 1966d[169]). Later, Garn *et al.* (1967j[115]) reported that the correlations for mesiodistal diameters between pairs of corresponding teeth in the upper and lower jaws were about 0.6. Correlations between teeth within a jaw for mesiodistal diameters showed the presence of canine and third molar fields and a field that included the more distal teeth of all classes (Garn *et al.*, 1963b[104], 1966d[169]). These findings indicated that the controlling factors could operate on part of the dentition.

The buccolingual diameter did not show systematic lateral differences, but, nevertheless, the differences tended to be larger for the more distal tooth in each morphological class (Garn, Lewis & Kerewsky, 1966e[110],

1967d[114]). These differences were slightly larger for boys than girls and they were largest for the upper second molar and smallest for the upper first premolar. There was marked similarity in the order of lateral differences for the two sexes (r = 0.8), but these lateral differences were not highly correlated with tooth size (r = 0.4). The lateral differences in the bucco-lingual diameters for various teeth were not significantly correlated within individuals and the lateral differences in mesiodistal and buccolingual diameters were not significantly correlated. Additionally, the rank order correlation between the sex differences in the mesiodistal and buccolingual diameters was low (0.2), due largely to the canines, which had the largest difference for the mesiodistal diameter, but the smallest for the bucco-lingual diameter (Garn et al., 1966d[169]).

Annual increments in the lengths of the mandibular canines, premolars, and second molars were reported that were based on highly reliable procedures (Israel, 1966; Israel & Lewis, 1971). Teeth with extremely curved roots were excluded and the mean root lengths were used for teeth with multiple roots. The plots of annual increments in length against age showed a wide scatter and slopes that differed by tooth within sex. The slope for the canine differed from those for the other teeth in males, but the slopes for the canines and premolars were similar in females. The slopes were similar for the two sexes, although the mandibular teeth were shorter in women than men because apical closure occurred earlier. In each sex, the premolar was shorter than the second molar from 6 to 8 years, but not at older ages.

Garn et al. (1966e[110]) found that the mesiodistal diameters of teeth did not tend to be larger on a particular side of the jaw, either in the total sample or in those with congenital absence of some teeth. Such findings may be of concern to orthodontists. The lateral differences in mesiodistal diameters had significant relationships with tooth size, and those for the maxillary teeth tended to be larger than those for the mandibular teeth. Furthermore, the more distal tooth of each morphological class (incisors, premolars, molars) tended to show more lateral asymmetry than other members of the class (Garn et al., 1966e[110]), and the positive correlations for lateral differences within individuals were larger for morphological classes of teeth than for the total dentition.

### Tooth shape

There is an interaction between the sizes of molar teeth and the number of cusps in different populations. To determine the nature of this association within the Fels population, Garn, Dahlberg and Kerewsky (1966f[75]) used the mesiodistal diameters of the first and second molars as an index of their size. The differences in size between these teeth were shown to be related to differences between them in the number of cusps (r = 0.4). Garn also defined tooth shape by the ratio buccolingual/mesiodistal diameter. There was a slight shape communality throughout the dentition (r = 0.2; Garn, Lewis & Kerewsky, 1967j[115]), with higher intercorrelations for adjacent teeth, especially those within the same morphological class. Furthermore, the ratio buccolingual/mesiodistal diameter was significantly larger in males, thus documenting sex differences in tooth shape (Garn *et al.*, 1967b[112]). These dimensions of teeth were moderately correlated (r = 0.5–0.6; Garn, Lewis & Kerewsky, 1968g[117]).

Garn *et al.* (1966f[75], 1966g[76], 1966h[77], 1966i[111]) also reported findings relevant to variations in the cusps and grooves on the crowns of molar teeth. Since they found greater similarity between brothers than sisters, and high concordance in monozygotic twins, they concluded that their findings were in reasonable agreement with genetic control by autosomal transmission. Furthermore, lateral differences were uncommon.

### Congenital absence of teeth

Congenital absence (agenesis) of the third molar teeth was shown to be related to the prevalence of absence of other teeth, and the timing of dental maturation and eruption (Garn & Lewis, 1962[100]; Garn, Lewis & Vicinus, 1963d[127]). Children with congenital absence of a third molar had a marked tendency to congenital absences of other teeth except for the first molar, and a reduction in the mesiodistal diameters of the other teeth, but children with a third molar did not exhibit congenital absence of any other tooth except the lateral incisors and second premolars (Garn *et al.*, 1962a[103], 1963b[104]).

When the third molar was congenitally absent, lateral differences in mesiodistal diameters of the other teeth were slightly larger than usual (Garn & Lewis, 1962[100], 1970[101]; Garn *et al.*, 1962a[103]; Garn, Lewis & Kerewsky, 1964a[105], 1966e[110]). This size reduction was more marked in the anterior than the posterior teeth. Additionally, when the distal tooth of one morphological class was absent, the corresponding teeth of other classes were likely to be absent (Garn & Lewis, 1962[100]; Garn *et al.*, 1963b[104]).

### Determinants of tooth size and shape

Genetic control over several aspects of dental size and shape has been documented at Fels by investigations based on correlations within families and the examination of pedigrees. There were significant correlations for mesiodistal diameters within sibling pairs that were larger for sister–sister pairs (0.6) than for brother–brother (0.4) or brother–sister pairs (0.2; Garn *et al.*, 1965a[107]). These genetic influences seemed to be more important than intrauterine pressure and other environmental influences. Additionally, profile patterns for mesiodistal diameters throughout the dentition were similar between siblings and parent–child pairs, and there was suggestive evidence of X-chromosomal involvement (Garn, Lewis & Walenga, 1968e[130]). The sex difference in mesiodistal diameters tended to be consistent within families. Some brother–sister pairs showed larger sex differences in mesiodistal diameters than did others; these sex differences were correlated about 0.5 between such pairs within families (Garn *et al.*, 1967k[125]). Additionally, data from siblings and twins showed that cusp number and cusp patterns were under genetic control (Garn *et al.*, 1966g[76]; Garn, Kerewsky & Lewis, 1966j[95]).

While definitive proof of a secular increase in tooth size was lacking, mesiodistal diameters tended to be larger in offspring than in their parents of the same sex (Garn, Lewis & Walenga, 1968c[129]). The increase was greater for males than females which would be expected because males are less influenced by genes on the X chromosome. This secular increase provided indirect evidence of environmental effects.

There were low (0.2) correlations between adult stature and the mesiodistal and buccolingual diameters of the permanent teeth (Garn & Lewis, 1958[99]; Garn, Lewis & Kerewsky, 1968d[116]). These low correlations led Garn and Lewis (1958[99]) to conclude that large fragmentary fossil teeth were not evidence that the prehistoric remains were those of giants. The sex differences in tooth size were correlated ($r = 0.3$) with the sex differences in stature within brother–sister pairs (Garn *et al.*, 1967c[113]).

# 7　Body composition and risk factors for cardiovascular disease

'It is not growing like a tree in bulk, doth make men better be.'

*Samuel Johnson* (1709–1784)

Interest in body composition at Fels is reflected in the data collection protocol from the first years of the study. This early interest was related to the amounts of adipose tissue, muscle and bone at local sites. The possible practical application of this work was soon realized. Garn (1962c[62]) wrote: 'Body composition attracts interest from clinical and preclinical disciplines,' but later (1963a[63]) he noted that, despite the established relationship of total body fatness to the probability of death (longevity) and of the amount of bone mineral to osteoporosis, little had been done to associate variations in regional body composition with the risk of disease or the probability of death. Garn (1963a[63]) also claimed that an excess of muscle mass predisposed to coronary atherosclerosis but tended to protect against osteoporosis.

The Fels Longitudinal Study was transformed in 1976 by placing an increased emphasis on serial changes in total and regional body composition in relation to risk factors for cardiovascular disease. The initial plan for the body composition aspect of the study was complex and further complicated by the irregularity of federal funding. As an overview, data are now collected annually from participants aged from 8 to 18 years who live within 80 km (50 miles) of Yellow Springs and then at 3-year intervals until 40 years, after which the examinations are at 2-year intervals. The intervals between examinations are longer for those who live further from Yellow Springs. The data collected at these examinations relate to the size, proportions, composition, maturity and density of the body, bioelectric impedance, blood lipids, selected hormones, and blood pressure. As a result, there are serial data for many important measures and near-unlimited analytic possibilities.

199

This chapter will consider work done within the Fels Longitudinal Study relating to: (i) methodological aspects, (ii) total body composition, (iii) bioelectric impedance, (iv) predictive equations, (v) regional body composition, (vi) adipocytes, (vii) the assessment of nutritional status, (viii) nutritional assessment in the elderly, (ix) influences on body composition, and (x) risk factors for cardiovascular disease.

### Methodological aspects
These methodological considerations will be restricted to work done to improve measurements of body composition. Research related to the interpretation, relationships, and applications of findings will be considered later in this chapter.

### *Total body composition*
The term 'total body composition' refers to the amounts of fat, fat-free mass, muscle, protein, bone mineral and water in the body. These amounts can be expressed using metric units or they can be expressed as percentages of body weight. Studies of bone mineral and skeletal mass are described in Chapter 6.

Many methods can provide body composition measures. Until recently, the most common laboratory procedure for the measurement of body composition was underwater weighing from which body density was obtained. Body density was then used to calculate the percentages of fat and fat-free mass in the body with the two-component model which assumed the densities of fat and fat-free mass were the same in all individuals.

The present method of choice is also based on the measurement of body density but the calculations of body composition are made using a multi-component model that takes account of variations in the composition of fat-free mass that affect its density (Garn, 1963a[63]; Roche, 1985a[199], 1987a[204]). From 7 to 25 years, age- and sex-specific group estimates for the density of fat-free mass can be applied; at older ages, total body bone mineral, total body water and body density must be measured to apply a multi-component model. There were large differences between values for body composition variables calculated from body density when the two-component model with a fixed value for the density of fat-free mass was used and those obtained from the current multi-component model (Lohman, 1986; Guo *et al.*, 1989a). These differences were larger in females than in males and they were larger at ages younger than 12 years when application of a fixed density for fat-free mass of 1.1 g/cc resulted in an underestimation of fat-free mass by about 2 kg.

The procedures now applied at Fels include the measurement of (i) total body bone mineral from photon absorptiometry using a dual-energy x-ray source, (ii) total body water from the dilution of deuterium oxide, and (iii) body density. This set of procedures allows the multi-component model to be applied on an individual basis, and thus provides excellent estimates for those old enough and fit enough to be weighed underwater. Furthermore, these data could lead to the development of improved multi-component models. Because these laboratory procedures require expensive fixed equipment, there is a need for effective field methods. These methods must be based on procedures that are easy to apply both to gather data and to use the data in predictive equations that provide accurate estimates of body composition values. The methods used to develop such predictive equations are described in Chapter 2; the efficacy of these equations will be described later in this chapter.

Garn and Nolan (1963[131]) described a tank for the measurement of body volume which, in combination with body weight, could be used to calculate body density. These workers chose a narrow tank to increase the accuracy of measurements of change in the water level when a participant was immersed. This level was read with an external vernier gauge to within 100 ml. The estimated combined errors of measurement of weight in air, residual volume, and body volume were considered to be more than 5%. Some of this error was due to the dimensional instability of the clear plastic tank.

In reviewing the progress of research concerning body composition, Roche (1984a[196]) referred to improvements in methods for developing predictive equations and the need either to discard the concept of frame size as a correlate of body composition or to develop an improved index of frame size. The present frame size measures explain little of the variance in percent body fat or fat-free mass that is not accounted for by weight or stature. The useful measures of frame size may be widths, such as knee width that is measured between bony landmarks over which there is little subcutaneous adipose tissue. Additionally, improved measurements of residual volume are required for the calculation of body density and, ideally, air in the gut would be measured. The variation introduced by the latter is reduced if the subjects are measured fasting (Roche, 1986c[203]). Later, Roche (1987a[204]) noted that load cells should be used to measure underwater weight; these measurements should be repeated ten times and the mean of the last three used in the calculations if these final measurements are representative of the total set. An alternative procedure is to use the mean of the three highest weights. Either of these means is a satisfactory choice but, of course, the choice must be fixed within a study and for comparisons between studies. The inherent limitations of the models are more important sources of error in body composition determinations than the errors of measurement which are small in good studies.

### Regional body composition

Attention to methods of measuring regional body composition at Fels have mainly been directed at subcutaneous adipose tissue. This is an important component of body composition, partly as an index of energy stores in the assessment of nutritional status (Himes, 1980a). Fat in adipose tissue is a highly labile energy store that accumulates due to a positive caloric balance. The amount of adipose tissue, however, is not a specific measure of caloric balance because it is affected by genetic influences, illness, malabsorption, parasitic infection, and habitual physical activity. The major rationale for the measurement of skinfold thicknesses is their close relationships to total body fatness.

Sites for the radiographic measurement of subcutaneous adipose tissue thickness must be chosen where the adipose tissue is sharply defined, standardized radiographic positioning is easy, and measurement errors are small (Reynolds, 1951). Slight variations in site location can have large effects on the recorded values. The best sites, such as the trochanteric, iliac crest and tenth rib, have underlying bony landmarks that allow accurate site location (Garn, 1957a[78]; Garn & Gorman, 1956[83]). Reliability was excellent for these sites and for the calf (Garn, 1961b[58]); measurements at them were highly correlated with weight and with corresponding thicknesses at other sites (Garn, 1954a[42], 1957b[49]). Measurement errors were large for sites that require oblique views such as the breasts and buttocks and for measurements over the abdomen where landmarks were lacking. Of course, if there is interest in a particular site, that is the one that should be measured. This interest may spring from known relationships between measures at the site and either total body fatness or risk factors for disease. At each site, the adipose tissue thickness is measured perpendicular to the skin surface and radiographic positioning must be such that pressure on the measurement site is avoided.

Subcutaneous adipose tissue thickness is also measured using calipers that exert the same pressure at all jaw openings. These measurements are limited to sites where the subcutaneous adipose tissue is loosely attached to the deep fascia and a double layer of skin and subcutaneous tissue (skinfold) can be elevated from the underlying muscle (Roche, 1979b[188]). It is commonly difficult to measure skinfold thicknesses over the abdomen, especially in the obese, because only mounds of tissue with non-parallel sides can be elevated. Harrison *et al.* (1988), describing methods for the measurement of skinfold thicknesses, emphasized the need to standardize the directions of the folds.

Himes, Roche and Siervogel (1979) compared caliper measurements of skinfold thicknesses compressed by calipers with radiographic measurements of uncompressed single adipose tissue thicknesses after correcting the latter for radiographic enlargement. These data for the analysis of the compressibility of skinfolds were obtained from participants aged from 8 to 19 years for seven sites, most of which were on the trunk. The median differences between half the skinfold values and the radiographic values would be zero in the absence of compression. This was almost the case for the calf and ulnar forearm sites but compressibility was present and varied markedly among the other sites. There were low ($r = 0.3$) correlations between sites for compressibility. In agreement with the findings of Garn (1956b[47]) for measurements over the tenth rib, compressibility in the data of Himes *et al.* (1979) did not change significantly with adipose tissue thickness. The sex differences in compressibility were not significant but there were significant differences in compressibility among sites, and significant interindividual variations for males but not for females. The differences in compressibility among sites and individuals may have been related to adipocyte size or the proportion of adipose tissue that was not fat (Roche, 1979b[188]). While it is important to recognize variations in skinfold compressibility, adjustments for this compressibility did not improve predictions of body fatness (Himes, 1980b).

In another study, Townsend (1978) compared the well-known Lange calipers with small plastic calipers (Adipometer®) developed and distributed by Ross Laboratories (Columbus, Ohio). With these plastic calipers, the pressure is exerted by the anthropometrist, not by a spring. Repeated bilateral measurements at the triceps and subscapular sites showed similar consistency for both types of calipers. The four plastic calipers were interchangeable for each site and measurements with them did not differ significantly from those with the Lange calipers. Nevertheless, as thicker skinfolds were considered, the differences between paired measurements (plastic–Lange) tended to change from positive to negative values and these differences were significant at both extremes of the distributions.

An alternative approach to the measurement of subcutaneous adipose tissue thickness employs ultrasonic waves. These waves do not compress the adipose tissue and, in theory, the method is applicable at sites where skinfolds cannot be measured. The ultrasonic waves are transmitted at right angles to the skin; in A-mode ultrasound the time required for their reflection from the surface of the underlying muscle to the skin is used to estimate the thickness of subcutaneous adipose tissue. Consequently, these measurements should be made at sites where the external surface of the underlying muscle is parallel to the skin (Roche *et al*,. 1985a[199]). Ultrasound can be used to measure the thickness of breast adipose tissue but site location is difficult, especially in the elderly. Despite the theoretical appeal of the method, data from A-mode ultrasound were shown to be inaccurate, in part because it was difficult to position the probe (transducer) so that its face was parallel to both the skin and the underlying muscle surface. Additionally, incorrect data were obtained when fibrous layers in the adipose tissue reflected the ultrasonic waves (Roche, 1979b[188]; Chumlea & Roche, 1986[27]; Roche *et al*., 1986a[217]).

In B-mode ultrasound, an image is obtained that can be stored as a permanent record on which tissue interfaces can be recognized. Therefore, the data are much more accurate than those from A-mode ultrasound. It is expected that B-mode ultrasound will be used more widely in the future despite its expense. B-mode measurements have recently been added to the examinations of Fels participants. Preliminary analyses of data from B-mode ultrasound show high reliability and close relationships to data obtained using calipers (Roche, Baumgartner & Siervogel, 1991b[214]). Considerable methodological work is in progress to evaluate this technique, with particular reference to deep abdominal adipose tissue.

### Total body composition
The investigations of total body composition at Fels will be considered in relation to weight and weight/stature$^2$, which are indices of total body composition, and to percent body fat, total body fat, and fat-free mass, which are basic components of body composition but are difficult to measure directly.

#### *Weight and weight/stature$^2$*
Body weight is highly relevant to body composition and it can be measured very accurately. It is important that measured values be used in the evaluation of either individuals or groups because reported data are inaccurate (Himes & Roche, 1982). Changes in adulthood have been documented. There were considerable increases in weight from 18 to about 30 years in men but the corresponding changes in women were small (Roche *et al*., 1975c[227]). In men and women, weight increased slightly from 19 to 70 years despite some decreases after 50 years that were more marked in men (Reynolds & Asakawa, 1950; Chumlea, Falls & Webb, 1982a[21]).

Although weight is the non-specific sum of the weights of all body components, it is useful because of its significant correlations (r = 0.8) with

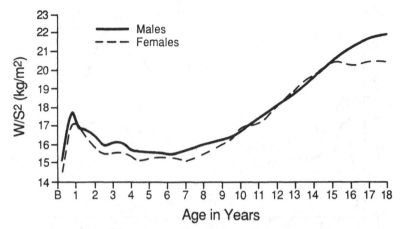

Fig. 7.1 Median values for weight/stature² plotted against age. (Cronk, C.E., Roche, A.F., Chumlea, W.C., Kent, R. & Berkey, C. (1982a). Longitudinal trends of weight/stature² in childhood in relation to adulthood body fat measures. *Human Biology*, **54**, 751–64, Redrawn with permission from *Human Biology*.)

total body fat and with the percentage of body weight that is fat (Roche *et al.*, 1981b[265]). Weight was also correlated with subcutaneous adipose tissue thicknesses at many sites; these correlations changed differently with age in boys and girls (Reynolds, 1951; Garn, 1961b[58]). In boys, the correlations increased from 7 to 11 years but decreased from 11 to 15 years. In girls, however, the corresponding correlations changed little from 7 to 11 years but increased from 11 to 15 years. In both sexes, there was considerable continuity of weight from one age to another after the age of 7 years, with correlations of 0.5 to 0.9 between values 5 to 6 years apart at ages ranging from 7 to 16 years (Reynolds, 1951).

Relative weight can be obtained by adjusting the observed weight to a median stature. This approach is not recommended, partly because comparisons between studies are difficult when the choice of reference data varies. Commonly, the weight of adults during middle age and old age is evaluated by comparison with data for weight at the same stature in young adults based on the assumption that increases in weight after young adulthood are due to increases in body fat and, therefore, undesirable. This view may be incorrect, especially for men in whom fat-free mass increases considerably after young adulthood (Roche, 1984c[198]).

The correlations between weight and body fatness were increased substantially when weight was divided by stature² to obtain the 'body mass index' (Roche *et al.*, 1981b[265], 1985[210]; Abdel-Malek, Mukherjee & Roche,

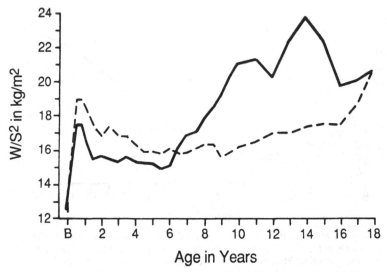

Fig. 7.2 Serial data for weight/stature$^2$ in two boys with markedly different scores for Principal Component 3 from 4 to 18 years. (Roche, A.F., Rogers, E. & Cronk, C.E. (1984) Serial analyses of fat-related variables. In *Human Growth and Development*, eds. J. Borms, R. Hauspie, A. Sand, C. Susanne & M. Hebbelinck, pp. 597–601[262], New York: Plenum Press. Redrawn with permission from Plenum Press, New York.)

1985). These correlations indicated that weight/stature$^2$ (W/S$^2$) may be useful in screening, but the standard errors of the predictions were about 3 kg for total body fat and 4% for percent body fat. Therefore, W/S$^2$ is not a useful predictor of body fatness.

Abdel-Malek *et al.* (1985) devised a weight–stature index with fractional powers, i.e., W$^m$/S$^k$ that were chosen to maximize the relationship between the index and percent body fat. These authors reported constants (coefficients) that, in combination with the new index, provided direct estimates of percent body fat with standard errors of 4–5%. The best index for Fels data was weight$^{1.2}$/ stature$^{3.3}$; the constants differed between the sexes but not with age. This index can be obtained easily from a published nomogram. This novel and logical approach to the development of an index of obesity was somewhat disappointing in that the chosen index was not significantly better than W/S$^2$ when judged by correlations with percent body fat. This work did, however, establish procedures that should be applied in the development of a weight–stature obesity index for non-whites or groups of abnormal individuals.

Serial changes in W/S$^2$ from 3 months to 18 years were described by Cronk *et al.* (1982a, 1982b). The medians increased rapidly to 1 year, then decreased to 6 years and subsequently increased until 18 years in males and 15 years in females (Fig. 7.1). Variations of $W/S^2$ within individuals were more marked during infancy and adolescence than in childhood, and there was less continuity from infancy to childhood than from childhood to adolescence and less continuity in females than in males. Additionally, it was shown that the mean percentile level for $W/S^2$ during an age range was the best single predictor of the value for $W/S^2$ at an older age. Age-to-age correlations were low to moderate until about 18 years in boys and 13 years in girls (Roche *et al.*, 1984[262]).

Using longitudinal principal component analysis of serial data for $W/S^2$ (Chapter 2), Cronk and her colleagues (1982b) assigned a coefficient to each component for every participant. Component 1 represented the average percentile level of the individual during the age range considered, while Component 2 represented the degree of directional change in percentile position (slope). Those with positive coefficients for Component 2 shifted upward with age relative to group percentiles. Components 3 through 6 represented smaller but increasingly more complex changes with age in $W/S^2$. These components described real aspects of individual variation in growth patterns as demonstrated by the data for two boys who differed markedly in their principal component coefficients from 4 to 18 years (Fig. 7.2).

There were strong relationships between the patterns of change in $W/S^2$ during growth and $W/S^2$ at 30 years (Cronk *et al.*, 1982a). For these analyses, data were analyzed for infancy (3 months to 3 years), childhood (4 to 18 years), pubescence (10 to 17 years), and young adulthood (18 to 30 years). In males, about half the variance in $W/S^2$ at 30 years was explained by the coefficients for Components 1, 2, and 6 for the age range 4 to 18 years; a similar fraction of the variance in females was explained by the coefficient for Components 1 and 2 for the age range 10 to 17 years. The components for changes during infancy explained little of the variance. Changes in weight from 18 to 30 years were correlated with the coefficients for Component 2 (3 to 9 years) and those for Components 1 and 5 (4 to 18 years) in males, and with the coefficients for Component 2 (3 months to 3 years) and those for Component 5 (4 to 18 years) in females. Adipocyte number in adulthood was related to the level of $W/S^2$ and the changes in $W/S^2$ during later childhood and adolescence but not the levels or changes during infancy (Cronk *et al.*, 1982a). The longitudinal principal components method described inflections but it did not provide clear descriptions of the variation in their timing, which is important (Siervogel *et al.*, 1989a). Mathematical functions have been fitted to serial data for $W/S^2$ from 2 to 18 years (Guo *et al.*, 1989c; Siervogel *et al.*, 1989a) that allow analyses of variation in the timing of inflections which is sometimes called the 'tempo' of growth. Since the fit was good, these models effectively described the patterns of change and allowed the calculation of the following for $W/S^2$: age at minimum value, the minimum value, age at maximum value, the maximum value, and maximum velocity. The maximum velocity and the maximum value occurred at younger ages in girls than in boys. Each of these estimated values, except age at maximum velocity, was related to the value of $W/S^2$ at 18 years, and early occurrence of the minimum value was related to high values for $W/S^2$ at 18 years.

There have been several analyses of the risk of having a weight of greater than the 75th percentile at 18 or 30 years, based on weight during childhood (Guo, 1988; Roche, 1987c[206]; Siervogel et al., in press) or of a high value for weight/stature[2] based on the percentile level of this variable in childhood. The probability of being overweight at 18 years was estimated from a logistic function that fitted the data well except for those below the tenth percentile. The risk differed little between percentile groups until 3 years and older. This demonstrated that excess weight in early childhood was not related to the risk of overweight in adulthood but there was a significant risk when excess weight occurred later in childhood. This risk increased with age and with the percentile level for weight or weight/stature[2] at the younger age. For corresponding weight groups, the risk of overweight in adulthood tended to be greater for females than males except after 14 years.

### Percent body fat and total body fat

In males, percent body fat calculated from the two-component model increased slightly early in adolescence but decreased by an average of 1.1% per year from 10 to 18 years; there was little change in females during the same age range and the age-to-age correlations were higher in females (Chumlea et al., 1981a[24], 1982b[38], 1983[39]). These conclusions may need revision when better multi-component models are available. Mukherjee and Roche (1984) fitted a quadratic function to serial predicted values for percent body fat during adolescence in each sex and showed marked variations in level but considerably less variation in the patterns of change. In men and women, percent body fat increased with age and the values tended to be higher in women at all ages (Chumlea et al., 1981b[33], 1982a[21]).

Total body fat calculated from body density did not change in boys during adolescence, but there was an average increase of 1.1 kg per year in girls (Chumlea et al., 1981a[24], 1982b[38], 1983[39]; Chumlea & Knittle, 1980[23]). The relationships between total body fat and maturity as measured by skeletal age were not significant in either sex (Chumlea, 1981[1]). Total body fat was greater in women than men from 20 to 24 years but, at older ages, the sex difference was not significant (Chumlea *et al.*, 1981b[33], 1982a[21]).

In early work at Fels, total body fat was predicted from trochanteric adipose tissue thickness (Garn & Harper, 1955[85]; Garn, 1957a[48], 1957d[51]). This approach yielded fat-free mass values that did not alter with changes in body weight, which is in conflict with many studies that show weight gains are associated with increases in both fat and lean tissues and that the reverse occurs when weight is lost. In fairness, Garn recognized that his approach was not applicable during periods of rapid weight change. This method may, however, be of value in screening but more data than a single skinfold thickness or radiographic adipose tissue thickness are needed for the accurate prediction of total body fatness.

### Fat-free mass

Little attention has been given at Fels to changes in fat-free mass with age, but Chumlea *et al.* (1983[39]) reported an average increase of 4.4 kg per year in boys from 10 to 18 years, with only slight changes in girls. Work related to fat-free mass has mainly concerned its prediction and its relationships to blood pressure, as will be described later in this chapter.

Muscle mass is an important part of fat-free mass but the measurement of muscle mass requires complex procedures based on measures of total body nitrogen and total body potassium. These methods were not applied in the Fels study but analyses were made of various indices of muscle mass. Muscle thicknesses were measured at selected sites on radiographs (Reynolds, 1951; Garn, 1961b[58], 1962b[61]), but these thicknesses are unlikely to be representative of total muscle mass, partly because the size of local muscle groups can be influenced by exercise. This view was supported by the near zero correlations between radiographic muscle thicknesses and visual ratings of 'muscle mass' (Garn, 1961b[58]). It was concluded, however, from local muscle thicknesses, that muscle mass increased by 5% from 17 to 35 years when the maximum was reached (Garn & Wagner, 1969[170]). Measurements of muscle thickness can be used to calculate muscle volumes for body segments, as has been done for the thigh muscle mass which is approximately cylindrical (Garn & Gorman, 1956[83]).

An index of muscle mass is provided by the amount of creatinine excreted daily in the urine. The mean values for creatinine excretion were similar in boys and girls until about 12 years when there were rapid increases in boys but not girls (Reynolds & Clark, 1947). Creatinine excretion had high day-to-day stability and its rate did not change during pregnancy (Seegers & Potgieter, 1937; Garn & Clark, 1955[72]). Despite the many sources of error in this index, Garn (1963b[64]) concluded that it was valuable from its correlation of about 0.7 with 17-keto-steroid excretion, stature, and calf muscle thickness.

Creatinine excretion corrected for body weight was proposed as an index of obesity because low values for the index reflected low values for muscle mass and fat-free mass relative to body weight (Garn & Clark, 1955[72]). The means for this index were larger in men than women and they were negatively correlated with adipose tissue thickness at the trochanteric site (r = −0.4). Nevertheless, this index was shown to be of limited value in men because of low sensitivity.

### Bioelectric impedance

Since the 1960s, it has been known that measures of the resistance of the body to the passage of a very small alternating electric current are influenced by the amount of body water. Water is a major constituent of the human body. Indeed, it has been remarked by Tom Robbins in *Another Roadside Attraction* that 'human beings were invented by water as a device for transporting itself from one place to another.' In a more serious vein, Saint-Exupéry described water as: 'Not necessary to life but rather life itself.' An estimate of total body water, in combination with selected anthropometric data, could assist the prediction of fat-free mass because all the body water is in the fat-free mass. This approach received a substantial boost in the early 1980s with the introduction of reliable electronic circuits and stable power sources. At the same time, considerable research was initiated to determine the extent to which this technique was useful, its underlying mechanisms, and its potential for surveys of nutritional status or body composition. The current commonly used is 800 $\mu$A with a frequency of 50 kHz. Pairs of source and receiving electrodes are applied to a hand–wrist and a foot–ankle.

The general topic of bioelectric impedance has been the subject of a critical review (Baumgartner, Chumlea & Roche, 1990). Resistance and reactance can be measured. Resistance is the opposition of the body to the flow of an alternating current. Capacitance, which is the reciprocal of reactance, causes the current to lag behind the voltage producing a phase shift that is quantified geometrically as the phase angle. At very low frequencies, the capacitance of the human body is effectively an open circuit (Fig. 7.3). Consequently, the reactance is zero and the measured impedance is purely resistive. With increasing frequency, reactance increases more rapidly than resistance and the phase angle enlarges until the critical frequency is reached. At frequencies higher than the critical level, reactance decreases relative to resistance until impedance is again

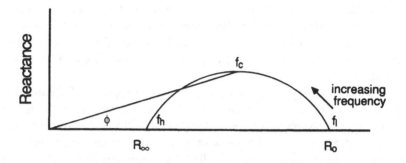

## Resistance

Fig. 7.3 An impedance plot that shows the relationship between resistance, reactance, and phase angle ($\phi$). The phase angle for a biological conductor is maximal at a critical frequency (fc). Very high frequencies are represented by fh where the capacitative component is essentially short circuited so that the measured impedance is purely resistive ($R\infty$). At low frequencies (fl) the capacitative component is effectively an open circuit: the reactance is equal to zero and the measured impedance is purely resistive. (Baumgartner, R.N., Chumlea, W.C. & Roche, A.F. (1988a). Bioelectric impedance phase angle and body composition. *American Journal of Clinical Nutrition*, **48**, 16–23. Copyright 1988. Redrawn with permission from *American Journal of Clinical Nutrition*.)

completely resistive at very high frequencies (Baumgartner *et al.*, 1990). In Fels data, the phase angle did not differ significantly between age and sex groups and it was not significantly correlated with fat-free mass in females, but there were low positive correlations in males (Baumgartner, Chumlea & Roche, 1988a).

### *Whole body impedance*
The application of bioelectric impedance to the measurement of body composition is based on the better conduction of electricity by fat-free mass than by fat, due to the high water content of fat-free mass. If the geometry of the human body were simple, few anthropometric data would be needed in combination with impedance to allow accurate predictions of body composition. Since the human body geometry is complex, whether one considers the whole body or its major conducting portion (water or fat-free mass), several anthropometric variables must be measured to assist the prediction of fat-free mass from resistance (Guo *et al.*, 1987a).

Measures of resistance were highly reliable and were not significantly related to the time of measurement within the range from 9 a.m. to 5 p.m., timing within the menstrual cycle, the use of oral contraceptives, or the interval from the previous meal or drink. Resistance was not associated with the level of habitual physical activity in adults, but women with a high level of physical activity had higher values for fat-free mass when estimated from resistance in combination with anthropometric values than when estimated from body density (Roche *et al.*, 1986a[217]; Chumlea *et al.*, 1987a[29]).

In work that was an important guide in the development of equations to predict fat-free mass from impedance and anthropometric values, it was shown that resistance was not correlated significantly with stature but had significant negative correlations with weight, circumferences, 'bone plus muscle' cross-sectional areas of the limbs, and skinfold thicknesses on the trunk (Baumgartner, Chumlea & Roche, 1987a). In combination, these anthropometric variables explained about 70% of the variance in resistance. Stature$^2$/resistance markedly improved the prediction of body composition in the absence of skinfold thicknesses but its contribution was markedly reduced when these thicknesses were included in the predictive model (Guo, 1986; Roche, Chumlea & Guo, 1987c[218]).

### Segmental impedance

Segmental impedance, which refers to the measurement of the resistance and reactance of body segments, is potentially important in the estimation of regional body composition, and in the estimation of total body composition for those in whom it is difficult or impossible to obtain impedance or anthropometric data for the whole body (Baumgartner *et al.*, 1990).

Resistance and reactance can be measured for large body segments, such as the trunk and limbs. Investigations of the possible usefulness of segmental impedance in the bedfast, in the prediction of total or segmental body composition, and in explaining underlying mechanisms are areas of active research in which Fels scientists are taking a leading part (Chumlea & Baumgartner, 1990[15]). These analyses showed that resistance and reactance were larger in women than men for the limbs, but the sex differences were not significant for the trunk. The same pattern of sex differences occurred in children except that the reactance of the trunk was larger in boys than girls (Baumgartner *et al.*, 1988).

For a uniform conductor, resistance is proportional to its length and inversely proportional to its cross-sectional area, i.e., $R = pL/A$ where L is the length of the conductor and A is its cross-sectional area. The coefficient, p, in this equation (specific resistivity) depends on the composition of the conductor. The resistivities for major parts of the body were higher for males than females and increased with age to 18 years after which there was little change (Baumgartner, Chumlea & Roche, 1989a;

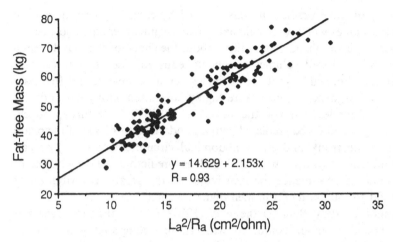

Fig. 7.4 The regression of fat-free mass (kg) on $La^2/Ra$ for the sexes combined. La = arm length, Ra = resistance of the arm. (Baumgartner, R.N., Chumlea, W.C. & Roche, A.F. (1989a). Estimation of body composition from bioelectric impedance of body segments. *American Journal of Clinical Nutrition*, 50, 221–6. Copyright 1989. Redrawn with permission from *American Journal of Clinical Nutrition*.)

Chumlea, Baumgartner & Roche, 1988a[18]). The resistivity of the trunk was about three times as great as that of the extremities in adults. This difference may be related to the direction of muscle fibers relative to the path of current flow. In general, the muscle fibers are parallel to the current in the extremities but not in the trunk, where resistivity may be influenced also by spaces that contain fluid or air.

Using this segmental approach, the sum of the predicted values for the fat-free mass in the arm, leg and trunk exceeded the whole body fat-free mass by a mean of 2 kg, but the values were highly correlated ($r = 0.9$; Chumlea, Baumgartner & Roche, 1986[16], 1987b[17]). These results would be assisted by better estimates of segmental volumes than those currently available from anthropometric data. Studies are planned in which segmental volumes will be measured from water displacement. Estimates of total body fat-free mass from the arm or trunk alone were reasonably accurate (Fig. 7.4). This major finding shows the possibility of using segmental impedance and anthropometry to predict body composition in those unable to stand (Baumgartner *et al.*, 1989b; Chumlea & Baumgartner, 1990[15]). This is important because few procedures are applicable to such individuals.

Chumlea *et al.* (1988a[18]) used the specific resistivity of muscle and estimates of muscle volume in the limbs to predict fat-free mass. The accuracy of these predictions was reduced by errors in the estimates of muscle volumes which were calculated from segment lengths and circumferences with simplistic assumptions about the shape of the muscle mass. Similarly, calculated volumes of subcutaneous adipose tissue in the arm and trunk, obtained by subtracting the segmental volumes for fat-free mass from total segment volumes were highly correlated with total body fat (r = 0.8; Chumlea, Baumgartner & Roche, 1988b[19]). With this approach, total body fat could be predicted with a standard error of 3 kg if the phase angle of the trunk and either abdominal circumference (men) or calf circumference (women) were included as predictors. The inclusion of abdominal circumference for men indicated the greater contribution of trunk fat to total body fat in men than in women.

In another study, Baumgartner *et al.* (1990) analyzed the effectiveness of total body resistance combined with stature compared with that of segmental resistances, combined with segmental lengths, in the prediction of fat-free mass in children. In the boys, data from the arm were the most effective but in the girls they were less effective than data from the whole body.

Investigations of impedance continue. Validation studies are in progress related to the prediction of total body water and the potential of impedance measures at multiple frequencies to estimate intracellular and extracellular water.

### Predictive equations

Fels scientists have been responsible for considerable progress in the development of predictive equations during the past 30 years. The most accurate methods for the measurement of body composition in living individuals require the use of sophisticated expensive equipment that is not portable and some require considerable cooperation from the subjects. These methods cannot be applied in field studies, or to large samples, or to individuals who are handicapped. Furthermore, methods that involve exposure to considerable amounts of radiation cannot be applied in serial studies due to the risks associated with cumulative effects. Therefore, simple procedures are needed that can provide data similar to what would be obtained by the complex laboratory methods. These simple procedures can be applied through predictive equations in which the independent variables are relatively easy to measure and the dependent variable is a body composition measure obtained with a complex laboratory method. The need for improved predictive equations was emphasized by Roche (1986c[203]) who noted that these equations should take account of variations in body water, skeletal mass, and, ideally, the sizes of some organs, especially the brain during infancy, and the intra-abdominal mass in adulthood.

Some predictive equations are considered indirect because they predict variables that can be used to calculate body composition measures. For example, Roche and Mukherjee (1982[254]) developed equations to predict body density from which body composition values can be calculated. These authors noted marked correlations among the predictor variables (multicollinearity) which would lead to instability of the regression coefficients in the predictive equations and poor performance when the equations are applied to other groups ('population-specificity'). To overcome this problem, they applied ridge regression in which a small constant (shrinkage parameter) was added to the diagonals of the intercorrelation matrix to reduce the interrelationships among the predictor variables and thus stabilize the regression coefficients.

In another analysis, Mukherjee and Roche (1984) used the maximum $R^2$ method to develop equations that predict percent body fat. This may be the first time this procedure was applied to body composition data. The maximum $R^2$ method was preferred to stepwise regression partly because the latter implies an order of importance for the predictors. In the maximum $R^2$ method, all the predictors selected early are retained and the best possible additions are made to the set of predictors. In these data, multicollinearity was not a major problem. The final equations explained about 75% of the variance, with a standard error of 4.0% in each sex. The predictor variables were age, abdominal circumference, triceps, and midaxillary skinfolds in each sex, in addition to age$^2$ and stature in men and biacromial width and lateral calf skinfold thickness in women. In replication groups, the errors were about 30% larger than those for the validation groups. The general levels and trends of serial values were similar for predicted and observed values of percent body fat (Fig. 7.5).

Other equations were developed to predict fat-free mass at ages from 7 to 25 years (Roche *et al.*, 1989b[241]; Guo *et al.*, 1989a). The outcome variable was calculated from body density by applying age- and sex-specific values for the density of fat-free mass. Fat-free mass was chosen as the outcome variable because the potential predictors included resistance from bioelectric impedance. As stated earlier, resistance is an index of total body water and all the body water is in the fat-free mass.

The statistical procedures used by Roche and Guo had been applied rarely, if at all, in a biological context. Descriptions of these procedures are provided by Guo *et al.* (1989a); here, only brief mentions will be made. All possible subsets regression was used to obtain the best sets of predictor (independent) variables for the present data. The best size set for the data

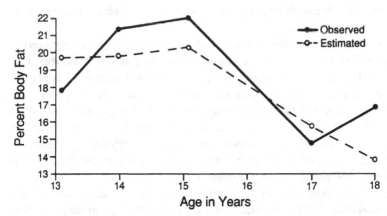

Fig. 7.5 A comparison between corresponding serial values for percent body fat estimated from an equation and calculated from body density ('observed') in one girl. (Mukherjee, D. & Roche, A.F. (1984). The estimation of percent body fat, body density and total body fat by maximum $R^2$ regression equations. *Human Biology,* **56,** 79–109, Redrawn with permission from *Human Biology.*)

was determined from $R^2$ and Cp values. Because the predictor variables might be highly intercorrelated, variance inflation factors were calculated. These variance inflation factors showed multicollinearity was a problem and, therefore, these correlations were reduced by ridge regression in combination with the PRESS procedure. Finally, to reduce the effects of individual values that have large effects on the regression coefficients, robust regression was used. As a result of these procedures in combination, it was considered that the best equation was selected and that its coefficients were modified to make it more likely to work well when applied to other groups.

The best equations for males and females had $R^2$ values of 0.98 and standard errors of 2.3 kg. The retained predictor variables were weight, lateral calf and midaxillary skinfold thicknesses, stature$^2$/resistance, and arm muscle circumference for males. The last-mentioned measurement is calculated from arm circumference and triceps skinfold thickness. For females, the retained predictor variables were weight, lateral calf, triceps and subscapular skinfolds and stature$^2$/resistance. Figure 7.6 shows the close correspondence between the predicted and observed values even at the extremes of the distribution.

Roche, Guo and Chumlea (1989d[240]) also developed equations that predict fat-free mass and percent body fat in adults. Resistance and reactance were 'forced in' as predictors and the selection of other predictor variables was determined by all possible subsets regression in combination

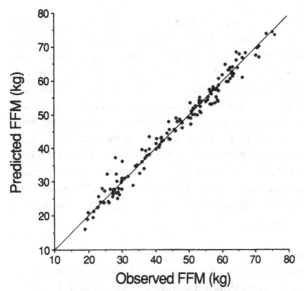

Fig. 7.6 The regression of values of fat-free mass (FFM) predicted from bioelectric impedance and anthropometry on 'observed' values from body density in males aged 7 to 25 years. (Guo, S., Roche, A.F. & Houtkooper, L. (1989a). Fat-free mass in children and young adults from bioelectric impedance and anthropometric variables. *American Journal of Clinical Nutrition*, **50**, 435–43. Copyright 1989. Redrawn with permission from *American Journal of Clinical Nutrition*.)

with the $R^2$ and Mallow's Cp procedures. The standard errors were about 2.5 kg for fat-free mass and 3.8% for percent body fat in each sex, which is extremely good since the errors in the outcome variables are probably about 1.8 kg for fat-free mass and 2.5% for percent body fat. Internal replication using a jackknife procedure yields closely similar results.

In another analysis, Roche and Guo (1990[238]) developed equations to predict fat-free mass in adults using all possible subsets regression with ordinary least squares to estimate the regression coefficients. The predictors were stature$^2$/resistance, reactance, reactance/resistance in each sex, with arm muscle area in men and calf muscle area in women.

The work done at Fels to develop equations that predict body composition variables from simple measures is far from complete. The data base enlarges week by week, new variables are being collected, and new statistical procedures applied. The most important of the latter is total body bone mineral which is now measured by dual photon absorptiometry which involves a very small amount of radiation and is, therefore, appropriate for serial studies.

### Regional body composition

There is considerable interest in regional body composition which concerns the distribution of tissues within the body. Much of this interest is related to well-established relationships between large amounts of adipose tissue over the trunk (truncal or central obesity) and the risk of some diseases, particularly non-insulin-dependent diabetes mellitus. Truncal obesity is a characteristic of men rather than women and may explain, in part, the greater risk of men for non-insulin-dependent diabetes mellitus and heart disease. As stated by Benjamin Franklin with a characteristically direct choice of words: 'Men run to belly and women to bum.' Garn (1963b[64]) lamented that 'human biologists interested in taxonomically important characters ignored fat because it is so obviously linked with the nutritional state.' This characteristic of adipose tissue makes it important for reasons not associated with taxonomy.

### *Adipose tissue*

As stated earlier in this chapter, local thicknesses of adipose tissue can be measured on radiographs, or by skinfold calipers, or ultrasound. From the early 1930s, following the lead of the Harvard Growth Study and the Child Research Council (Denver), radiographs were taken of Fels participants to allow the measurement of tissue thicknesses at standardized sites. The radiographic approach requires considerable attention to detail; even then, the error of measurement is about 4%. The recorded data should be corrected for radiographic enlargement, but commonly this was not done. While the radiographic method has many advantages, the associated irradiation is a serious problem, although the skin dose can be very low and the gonadal dose can be near zero (Garn, 1961b[58]).

Garn (1957d[51]) reported that in young men, the correlations between radiographic adipose tissue thicknesses at various sites were generally highest for the deltoid, tenth rib, iliac crest and trochanteric sites. Due to these findings, and because the thicknesses at these sites were most highly correlated with weight, Garn selected them as the most useful for the assessment of nutritional status (Garn, 1954a[42], 1957a[48], 1957b[49], 1957c[50], 1957d[51]; Garn & Clark, 1955[72]). Later, Roche (1979b[188], 1979c[189]; Roche *et al.*, 1985[210]), mainly on the basis of practicality and relationships to percent body fat reported by Siervogel *et al.* (1982b), recommended the triceps, anterior chest and subscapular skinfold sites for children. Roche further recommended that measurements at the anterior chest and triceps skinfold sites be included for the assessment of nutritional status in young men and women, that the midaxillary and paraumbilical sites be included for men, and the subscapular and supra-iliac sites for women. For older men, he recommended the subscapular site and for older women the anterior chest, paraumbilical, subscapular and chin sites.

### Age changes in adipose tissue thicknesses

A large effort has been expended at Fels to describe changes with age in thicknesses of adipose tissue. It was shown that during pubescence, adipose tissue thicknesses decreased in males but increased in females. These thicknesses tended to be larger and more variable for girls than for boys and they were about twice as large in women as in men (Reynolds, 1948, 1951; Reynolds & Grote, 1948). Reynolds (1946a) reported that rapidly maturing girls had larger calf adipose tissue thicknesses and larger relative gains than slowly maturing girls but less variability. Rapidly and slowly maturing boys differed in a similar way but to a lesser extent. Garn and Haskell (1959b[87]) showed that adipose tissue thicknesses over the tenth rib increased from 6 to 18 years in an almost linear fashion in girls but there were only small changes in boys during the same period.

Garn (1954a[42], 1957b[49]) reported that adipose tissue thicknesses tended to be greater in women than men except at the iliac crest and deltoid sites where the thicknesses for men were slightly greater. Additionally, Garn (1960c[56]) reported large increases from 20 to 40 years in males at the iliac crest and in females at the trochanteric site.

Reynolds (1949b) constructed a ratio of adipose tissue/total bone thicknesses for the calf which he proposed as a sex-differentiating character. This index had an error of about 30% in sex determination at 7 to 13 years, but the error was about 6% during adolescence and was 4% for men and 16% for women. Adipose tissue thicknesses relative to the total width of the calf were also greater in females than males at all ages (Reynolds, 1948; Reynolds & Grote, 1948).

In an extraordinarily complete monograph, Reynolds (1951) reported radiographic thicknesses of adipose tissue at various sites from 6 to 17 years and he analyzed data for infants and adults including some unusual sites. On the medial aspect of the thigh, measurements were made at the level of the crotch, and where the adipose tissue thickness is maximal just proximal to the knee. Both these adipose tissue thicknesses were larger in girls than boys. Thicknesses on the lateral aspect of the thigh at the level of the crotch were considerably larger in girls than boys from 6 to 10 years although the increases were larger in boys than girls during this age range. Reynolds also reported that adipose tissue at the base of the neck tended to be larger in girls than in boys but changed only slightly from 6 to 10 years.

The median thicknesses decreased before pubescence in boys, especially

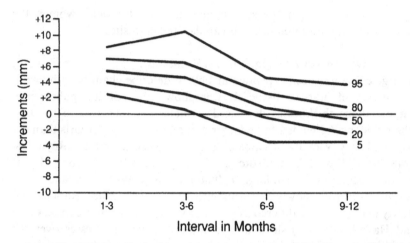

Fig. 7.7 Percentiles for increments in the sums of medial and lateral calf adipose tissue thicknesses in girls during infancy. (Garn, S.M., Greaney, G. & Young, R. (1956d[84]). Fat thickness and growth progress during infancy. *Human Biology*, **28**, 232–50. Redrawn with permission from *Human Biology*.)

at the waist; this decrease ceased earlier at the tenth rib site than at other sites. When expressed relative to the sum of the adipose tissue thicknesses for the six sites measured, the relative thickness of the calf adipose tissue decreased linearly with age in each sex from 8 to 16 years (Reynolds, 1951). In girls, there was a linear relative increase at the trochanteric site to 18 years, but the increase in boys stopped at 12.5 years and was followed by a decrease. The relative thickness at the tenth rib site increased linearly but at a slow rate in each sex, while that at the forearm site decreased linearly in each sex. In both boys and girls, there were slow linear decreases in relative adipose tissue thickness at the deltoid site after 8 years. At the waist site, the relative thickness increased linearly in boys to 18 years but the increase stopped at 13 years in girls (Reynolds, 1951). This was in agreement with other data showing the early development of truncal fat distribution in males (Roche & Baumgartner, 1988[211]).

There were only small sex differences in the medians but wider distributions in females than males for the sum of medial and lateral calf adipose tissue thicknesses in infants (Garn, 1956b[47]). In each sex, the sums increased rapidly to 6 months, then slowly to 9 months and decreased slightly from then to 12 months. While the group data indicated only slight changes after 6 months, there were large increments in many infants but they differed in direction. This phenomenon is illustrated in Fig. 7.7, which shows negative increments were increasingly common after 3 months.

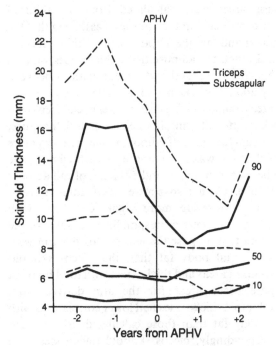

Fig. 7.8 Percentiles for triceps and subscapular adipose tissue thicknesses (mm) in relation to age at peak height velocity (APHV) in boys. (Cronk, C.E., Mukherjee, D. & Roche, A.F. (1983b). Changes in triceps and subscapular skinfold thickness during adolescence. *Human Biology*, **55**, 707–21, Redrawn with permission from *Human Biology*.)

Cronk, Mukherjee and Roche (1983b) calculated selected percentiles for skinfold thicknesses grouped at ages relative to the timing of peak height velocity (age at peak height velocity, APHV). In boys, triceps skinfold thicknesses increased prior to PHV but decreased near the time of PHV. The corresponding values for girls increased prior to PHV, declined near this event except at the 90th percentile level, and increased later. The percentiles for subscapular skinfold thicknesses in boys and girls showed only slight changes at about the time of PHV, except at the 90th percentile level where there was an increase followed by a decrease (Fig. 7.8).

### *Adipose tissue areas*
The calculation of adipose tissue areas in cross-sections of the limbs has received the attention of Fels scientists for many years. Garn (1961b[59]) and Roche (1979b[188], 1979c[189], 1986c[203]) recognized the limited validity of cross-sectional tissue areas but accepted their conceptual appeal and

potential usefulness. These areas were calculated from the skinfold thickness at a site and the circumference at the same level assuming that the total cross-section was circular and that the adipose tissue within it formed an annulus of fixed width. In fact, the adipose tissue thicknesses vary on different aspects of a limb at the same level and the correlations between these thicknesses were only about 0.6 (Garn, 1954a[42], 1957c[50]).

These adipose tissue areas could distinguish between children with equivalent circumferences but different skinfold thicknesses and, therefore, different areas of adipose tissue. Indeed, calf adipose tissue area was more highly correlated with weight than was calf adipose tissue thickness; the correlations were even higher between weight and estimates of calf adipose tissue volume (Garn, 1961b[58]). Garn also showed that adipose tissue thicknesses and areas of the arm were more highly correlated with measures of total body fatness than were corresponding measures of the calf. Himes *et al.* (1980) found that adipose tissue areas for the arm were more highly correlated with total body fat than the triceps skinfold thicknesses but the correlations of areas and thicknesses with total body fat were about equal. Adipose tissue areas for the arm derived from circumferences and biceps skinfold thicknesses had lower correlations with total body fat or percent body fat than those calculated using triceps skinfold thicknesses. Correspondingly, biceps skinfold thicknesses were less closely correlated with total body fat or percent body fat than were triceps skinfold thicknesses.

### Distribution of adipose tissue

The potential importance of the distribution of adipose tissue was recognized early at Fels. Garn and Harper (1955[85]) wrote: 'It is not unlikely that the location of the fat, as well as the gross amount, may elucidate the differential mortality rates of different weight groups.' The topic of adipose tissue distribution was, however, bedeviled by uncertainty about definitions and methods of measurement (Roche, 1986c[203]) despite the significant methodological contributions made by Garn (1955b[45]) and by Baumgartner *et al.* (1986b). The usual approach was to contrast measures of adipose tissue on the upper and lower parts of the body or measures on the trunk with those on the extremities (Reynolds, 1951; Garn, 1954a[42], 1955b[45]). The measures of adipose tissue may be radiographic thicknesses, skinfold thicknesses, or circumferences. Indices based on circumferences, such as the waist–hip ratio, are not specific to subcutaneous adipose tissue because they are affected by muscle, bone and deep adipose tissue. Nevertheless, they are commonly regarded as indices of adipose tissue distribution. Logarithms may be utilized and measurements may be combined as an index or one value may be regressed against another. When there are measurements at three or more sites, principal components analysis has been used to define indices of adipose tissue distribution.

Garn (1955a[44], 1955b[45]) converted adipose tissue thicknesses at nine sites to standard scores and used these to describe profiles for adults. The standard deviations of the standard scores for individuals were regarded as measures of variability. In men, variability was not related to age or the mean adipose tissue thickness but it increased when the adipose tissue was relatively thick at the trochanteric site. This could indicate that the thickness at this site was relatively independent of the thicknesses at the other sites, but the trochanteric site had the highest mean correlation coefficient (communality index) for men and the iliac crest had the highest for women.

Garn (1954a[42]) reported an association between the distribution of adipose tissue and body weight. In men weighing less than 74 kg, the largest mean adipose tissue thickness was at the deltoid site, but in heavier men the trochanteric thickness was the largest. Later, Garn (1955b[45]), in agreement with Reynolds (1951), considered that the pattern of adipose tissue distribution was independent of body fatness.

A large literature relates the ratio between hip and waist circumferences to the risk and presence of some chronic metabolic diseases. These interesting publications have led to conjecture about which of the tissues included in these circumferences are responsible for the relationships that have been noted. There is little relevant information but, Reynolds (1951) reported correlations of about 0.7 between radiographic trochanteric adipose tissue thickness and bi-trochanteric width measured without pressure. The corresponding correlations between this adipose tissue thickness and bi-iliac width, which is essentially a skeletal measure recorded with pressure, was only 0.4. Furthermore, Reynolds (1951) reported only moderate correlations (r = 0.5) between pairs of adipose tissue thicknesses and circumferences at the same level for the calf, upper arm, and waist.

These findings indicate that a circumference measurement is influenced by tissues (adipose tissue, muscle, and bone) that are not highly correlated. Therefore, ratios between circumferences may lack specificity regarding adipose tissue distribution. This has been demonstrated using hospital records from computed tomography (Baumgartner *et al.*, 1988b). It was shown that the ratio of abdominal to hip cross-sectional areas was correlated strongly with intra-abdominal fat but also with muscle and bone areas. After adjustments for body fatness, the ratio between waist and hip circumferences was only 0.3 to 0.4.

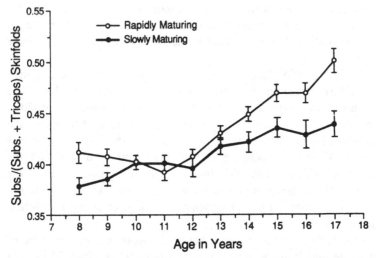

Fig. 7.9 Changes with chronological age in the ratio (subscapular/ subscapular
+ triceps skinfold thickness) for rapidly and slowly maturing boys. (Xi, H.,
Roche, A.F. & Baumgartner, R.N. (1989a). Association of adipose tissue
distribution with relative skeletal age in boys: The Fels Longitudinal Study.
*American Journal of Human Biology*, 1, 589–96, Copyright 1989. Redrawn with
permission from Wiley – Liss.)

Investigating a novel aspect of body composition, Garn *et al.* (1954[163])
showed that baldness was not associated with scalp thickness. Scalp
thicknesses, which are largely determined by adipose tissue, measured on
lateral radiographs, increased rapidly from 4 to 15 years in each sex and
then increased slowly to 30 years.

### Patterns of change in adipose tissue distribution
The distribution of adipose tissue in childhood is of interest because of the
possible tracking of patterns from childhood to adulthood. Garn documented the concurrent gain of adipose tissue thickness at some sites and
loss at others for adults (Garn & Young, 1956[172]; Garn, 1956a[46]) but little
information is available about the tracking of adipose tissue distribution
because few serial studies have included measurements of adipose tissue
thickness at multiple sites. Some tentative conclusions are possible from
Fels reports.

Reynolds (1951) reported serial data for a few individuals. In a boy, the
changes at the forearm site differed from those at the other sites but there
was considerable parallelism (tracking) between sites in the girl whose data
were displayed in his monograph. Correlations between ratios of skinfold
thicknesses across 5-year intervals, beginning at 7 to 13 years, were about

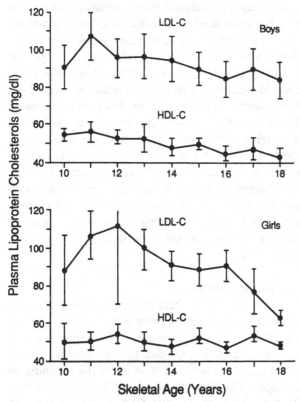

Fig. 7.10 Means for low-density lipoprotein-cholesterol (LDL-C) and high-density lipoprotein-cholesterol (HDL-C) plotted in relation to skeletal age. (Siervogel, R.M., Baumgartner, R.N., Roche, A.F., Chumlea, W.C. & Glueck, C.J. (1989b). Maturity and its relationship to plasma lipid and lipoprotein levels in adolescents. The Fels Longitudinal Study. *American Journal of Human Biology*, **1**, 217–26. Copyright 1989. Redrawn with permission from Wiley – Liss.)

0.7. The correlations tended to be higher for boys than girls and increased significantly with age in the boys but not the girls (Roche & Baumgartner, 1988[211]). From 11 to 18 years, particularly in boys, changes in indices of adipose tissue distribution reflected an increasing relative concentration of adipose tissue on the trunk (Baumgartner *et al.*, 1986b).

Xi, Roche and Baumgartner (1989a) analyzed the relationship between the distribution of adipose tissue and relative skeletal age (skeletal age less chronological age) in boys. This was prompted by the possibility that pubescent changes in adipose tissue distribution might occur earlier in rapidly maturing boys than in slowly maturing boys. These workers showed that after 13 years, rapid maturation was associated with a more truncal pattern (Fig. 7.9) but the rate of maturation from 7 to 14 years did not predict the adipose tissue distribution at 17 years, when adjustments were made for baseline values. Relationships between skeletal age and plasma lipids have been analyzed (Siervogel *et al.*, 1986b; Fig. 7.10). These analyses showed that low-density lipoprotein-cholesterol (LDL-C) changed little with increasing skeletal age in boys but decreased markedly in girls. The values for high-density lipoprotein-cholesterol (HDL-C) changed only slightly with increasing skeletal age in each sex.

Despite more than 50 years of effort, much remains unknown in relation to regional body composition. To some extent, this is due to the limitations of the methods. Particular areas of interest include tracking, the measurement of superficial adipose tissue with B-mode ultrasound and of deep adipose tissue with magnetic resonance imaging or ultrasound, and the measurement of segmental body composition by dual photon absorptiometry and bioelectric impedance.

### *Tracking of fat-related variables*
In addition to studying the differences between means or medians of adipose tissue thicknesses at successive ages, there is considerable interest in 'tracking.' Tracking refers to the maintenance of rank order within a group of individuals over time. This can lead to the identification of childhood antecedents of adult obesity and thus assist the planning of effective preventive strategies. The simplest measure of tracking is an age-to-age correlation. Tracking can be judged also from the degree to which well-fitted curves for individuals are parallel within a specific time period. Some consider 'tracking' can also refer to 'canalization' which is the tendency of a trait to proceed to a predictable endpoint. Canalization is best judged by fitting a mathematical model to the serial data for an individual; the extent of canalization can be judged by the goodness of fit of the model.

In an early Fels study, it was shown that adipose tissue thickness at the tenth rib site at birth was correlated ($r = 0.5$) with calf adipose tissue thickness at 1 month (Garn, 1958a[52]). There was also a low negative correlation ($r = -0.3$) between the adipose tissue thickness at the tenth rib site and positive correlations in weight gain from birth to 3 months and both weight ($r = 0.3$) and recumbent length ($r = 0.2$) at 3 months. These studies did not directly concern tracking since the variables were different at the two ages but they demonstrated slight continuity of growth.

Garn (1956a[46], 1962b[61]) separated infants into quartiles for total calf adipose tissue thickness at 1 month. The initially 'fat' infants were larger

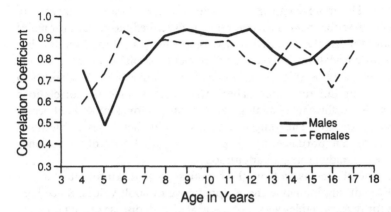

Fig. 7.11 Age-to-age correlations across 1-year intervals for triceps skinfold thicknesses. (Roche, A.F., Siervogel, R.M., Chumlea, W.C., Reed, R.B., Valadian, I., Eichorn, D. & McCammon, R.W. (1982a). Serial changes in subcutaneous fat thicknesses of children and adults. *Monographs in Paediatrics*, 17. Karger, Basel[264]. Redrawn with permission from Karger, A.G., Basel.)

and had larger calf muscle thicknesses than those in the lowest quartile at 1 month but these differences were not significant at 6 months. Reynolds (1951) reported moderate correlations ($r = 0.7$) between adipose tissue thicknesses 5 years apart for various sites and for girls and boys, except from 8 to 13 years when the correlations for boys tended to be lower. Additionally, Reynolds reported close relationships between the sums of adipose tissue thicknesses at 8 years and at menarche in girls and for values at corresponding ages in boys. In an earlier study, Reynolds (1944) reported that children typically retained their patterns of tissue components in the calf from 7 to 10 years.

Age-to-age correlations for skinfold thicknesses and $W/S^2$ increased markedly to 6 years for most sites on the arm and trunk but not on the calf (Roche *et al.*, 1982a[264]; Roche, Rogers & Cronk, 1984[262]). Data for the triceps skinfold thickness are shown in Fig. 7.11. These age-to-age correlations decreased during pubescence for only a few sites. Correlations between values at younger ages and those at 16 years or in adulthood increased rapidly to 6 years for most sites. The levels of the correlations were similar in both sexes and were generally similar on the trunk and the extremities. The correlations for radiographic data were consistently higher than those for skinfold thicknesses. This may have reflected the smaller errors of measurement.

Cronk *et al.* (1983a) analyzed serial skinfold thicknesses using longitudinal principal components to summarize significant patterns of change with age. The more complex these patterns, the more components are needed to describe them. Two components described the pattern for males but those for females required four components. There were modest but significant correlations between the coefficients of analogous components for different skinfold thicknesses, indicating a tendency for the changes to be similar at different sites. There were also significant correlations between the coefficients of analogous components for skinfold thicknesses and weight/stature$^2$. The components for skinfold thicknesses in childhood had long-term importance; they explained about 20–50% of the variance in indices of body fatness in adulthood.

The general conclusion from these analyses is that skinfold thicknesses and $W/S^2$ during infancy were not predictive of adult values. Since high values for these variables were not associated with disease during infancy, obesity during the first few years after birth is not a medical problem (Roche, 1979b[188]). The proverbial wisdom of Sei Shonagon can be accepted: 'Small children and babies ought to be plump.' Nevertheless, there was significant tracking to adult values after 4 years, despite which, childhood skinfold thicknesses were not closely related to the changes in the same variables from 18 to 30 years.

### Muscle

Muscle thicknesses in the calf, measured on radiographs, changed similarly with age in each sex from birth until pubescence, when there were much more rapid increases in males than in females, particularly in rapidly maturing children (Reynolds, 1944, 1948; Reynolds & Grote, 1948). Muscle thicknesses continued to be greater in males than females throughout adulthood (Garn & Saalberg, 1953[159]). Cross-sectional areas of adipose tissue and 'muscle plus bone' in the arm and calf decreased with aging, which indicated decreases in the fat-free mass and fat of the extremities (Chumlea, Roche & Mukherjee, 1986[31]). At ages from birth to 6 years, there were only low correlations between the thicknesses of muscle and adipose tissue in the calf and both of these thicknesses had only low correlations with total bone width (Reynolds, 1944, 1951; Reynolds & Asakawa, 1950; Garn, 1956a[46]).

Reynolds and Asakawa (1950) found associations between unusual tissue distributions in the calf and unusual body shapes in adults. Using terms derived from somatotype studies, they defined endomorphs as individuals with rounded contours and large circumferences, especially of the abdomen, while mesomorphs were defined as individuals with a large frame and well-developed muscles. Ectomorphs were recognized by their slender build. In the Fels group, endomorphs had wider calves with relatively more adipose tissue, while mesomorphs had calves of inter-mediate width with large thicknesses of bone and muscle. Each of these tissues had low values in ectomorphs. Some behavioral associations were reported by Garn (1962b[61], 1963b[64]) who found that muscle thickness in the calf at 6 months was predictive of the age at which an infant will first stand, crawl and walk.

### Adipocytes

The cells in adipose tissue that store fat, mostly triglycerides, are known as adipocytes. There is considerable interest in these cells because the amount of body fat must approximate the product of the number of adipocytes and the mean weight of these cells. In addition, there are small amounts of fat in body fluids, cell membranes and neurological cells. When simple procedures became available to measure the weight of adipocytes using osmium tetroxide, there was a marked increase in research relating to them. Some of this research was done at Fels although estimates of the weight and number of adipocytes are limited in accuracy (Roche, 1979b[188]). Adipocyte weight can be measured reliably but in small samples that are not representative (Chumlea *et al.*, 1981a[24], 1981b[33]). The small samples lead to errors when adipocyte number is calculated by dividing total body fat by mean adipocyte weight. This subject is further confounded by the presence of cells (pre-adipocytes) that lack fat and, therefore, are not measured, but can mature into adipocytes that store fat.

In critical reviews, Roche (1979b[188], 1981b[194]) evaluated the hypothesis that obesity during infancy is associated with an increased number of adipocytes that persists throughout life and provides a mechanism for the development and persistence of obesity. The near zero correlations between measures of body fatness in infancy and adulthood are not in agreement with the view that adult obesity is determined by the number of adipocytes present in infancy (Roche *et al.*, 1982a[264]). In addition, the estimated number of adipocytes can change in adults, and rapid increases in total body fat during infancy may reflect increases in adipocyte weight, number, or both.

Chumlea and his colleagues (1981a[24]) reported that, from 10 to 18 years, girls had larger gluteal adipocytes than boys and larger estimates of total adipocyte number. During this age range, adipocyte weight decreased significantly with age in boys ($r = -0.4$) but not girls and adiopocyte number increased in both sexes. These findings were not in agreement with the hypothesis that adipocyte number is fixed early in life. But during childhood, age-to-age correlations across annual intervals were significant for adipocyte number but not for adipocyte weight. Adipocyte number increased slightly with age in men and women and, as expected, gluteal adipocytes were heavier in women than in men (Chumlea & Knittle, 1980[23]; Chumlea et al., 1981b[33]).

These findings for gluteal adipocytes can be generalized to some extent because there are high correlations among adipocyte weights at various sites in children. In addition, gluteal adipocyte weight is more highly correlated with total body fat than is adipocyte weight at other sites. Nevertheless, the use of a single site and the inability of the osmium tetroxide method to detect adipocytes containing little or no fat lead to the need for caution in any generalization from these results (Chumlea et al., 1982b[38]).

After adjusting for age, there was a small but significant ($r = 0.3$) correlation between blood pressure and adipocyte number in each sex, but the correlations with adipocyte weight were not significant. Furthermore, total body fat was not correlated with gluteal adipocyte weight, so differences in total body fat appeared to be associated with gluteal adipocyte number (Siervogel et al., 1982c).

Further analyses of data for adipocyte weight are in progress that relate to values at an age, patterns of change, tracking, and associations with risk factors.

### Assessment of nutritional status

Nutritional status can be assessed for an individual or a group and it has both static and dynamic connotations. Nutritional status can be determined completely only by measuring the circulating levels and body stores of all nutrients. A complete set of direct measurements would require many organ biopsies and fluid samples and a multitude of chemical analyses. Such an assessment is clearly impractical and its effectiveness would be limited by the inaccuracy of many laboratory methods and the rapid changes that occur in some of these measures. Work at Fels concerning the assessment of nutritional status has been limited to body stores of fat and bone mineral and to the measurement of bone, adipose and muscle tissues, and fat-free mass. The scope of this work has been wide in that procedures have been developed to predict and interpret these values in field studies of large samples and reference data have been provided (Malina & Roche, 1983). This account relates to the application of indices of nutritional status and of equations that predict body composition variables from data that are easy to obtain without invasive procedures.

The need for such methods was stressed in a Wenner–Gren Conference on the Measurement of Nutritional Status (Roche & Falkner, 1974[231]). It was recommended that field measures should be simple, cheap, and applicable by observers who need only a moderate amount of training. The instruments should be light and the measurements must be acceptable to the population to be studied. Anthropometric variables are almost always included in a list of measurements for the assessment of nutritional status in a field study. These measurements, in common with all others, must be made accurately and be capable of interpretation. For example, a deficit in head circumference at 2 years may indicate malnutrition during infancy; such inferences regarding the timing of malnutrition may be improved by the concurrent evaluation of weight, recumbent length, and weight-for-length. These inferences should be made cautiously, especially for individuals, since genetic influences may be involved.

The findings of Roche *et al.* (1981b[265]) are relevant if the assessment of nutritional status is oriented towards body fat. These authors considered the grading of body fatness when only limited anthropometric data were available. This work was undertaken because the associations between common measurements and body fatness were not well-established and, therefore, the interpretation of these measurements was uncertain. Using data from 6 to 49 years, they compared the relationships between several measured and derived variables and percent body fat and total body fat. They concluded that the triceps skinfold was the best single index of percent body fat in children and women, and the best index for men was $W/S^2$. These authors suggested that these measures be compared with national reference data in screening for body fatness. To facilitate this, a nomogram was published that provided $W/S^2$ from the measured variables.

In the assessment of nutritional status, observed anthropometric values may be adjusted, such as the adjustment of weight for stature, or they can be used in predictive equations to estimate body composition variables (Roche, Baumgartner & Guo, 1987a[212]; Roche & Chumlea, in press[216]). The latter procedure should be applied only after the equations have been satisfactorily replicated on a subset of the study population or one closely similar to it. Additionally, as explained in Chapter 2, it may be desirable to transform observed variables to remove skewness or to calculate areas, ratios or other indices from them.

Roche (1974d[181], 1982[195], 1989b[208]) and Roche *et al.* (1985[210]) warned against the ready acceptance of the sensitivity of $W/S^2$ because its relationship with body fatness could differ markedly among populations. Indices of body fatness derived from weight and stature need not be independent of stature as has been claimed by some. Indeed, this characteristic would make an index undesirable during childhood, when fatness is related to stature (Himes & Roche, 1986). The best basis on which to judge such indices is the proportion of the variance in body fatness for which they account.

Attention was directed to the importance of the ratio head circumference/chest circumference which changes with age due to the different rates of growth of these body parts. This ratio is a guide to long-term malnutrition during the pre-school years. Additionally, weight-for-length, weight-for-stature and arm circumference are useful in rapid screening, particularly in developing countries. Due to their essential independence from age in early childhood, age need not be known to interpret the values. Robinow (1968) suggested that low values for arm circumference were due to decreases in 'muscle plus bone' rather than adipose tissue and, therefore, they are guides to protein reserves.

Reynolds and Asakawa (1948) presented a tentative set of screening procedures relating to body fatness using bivariate relationships between weight, stature, and calf adipose tissue thickness. This interesting approach was not developed to the stage at which it could be applied. Later, Garn (1962c[62]) suggested the use of weight adjusted for radiographic chest breadth because weight was highly correlated with chest breadth.

Following Roche (1984c[198]), Chumlea (1985b[6], 1985c[7], 1986a[8], 1986b[9]) pointed out that total body fat and percent body fat differ in their social acceptability during childhood and in their associations with disease in adulthood. Consequently, the assessment of nutritional status should take into account both total body fat and percent body fat.

These authors reviewed the anthropometric approach to the assessment of nutritional status, including total body fat and percent body fat, in children with developmental disabilities. Excessive curvature of the vertebral column may interfere with the measurement of stature and complicate its interpretation. In such children, arm span can replace stature, with an adjustment for the expected differences between these measures, but in many disabled children arm span cannot be measured accurately. Some disabled children must be measured in unusual body positions and there is a lack of disease-specific reference data with which values can be compared. This lack is difficult to overcome because data are needed from large groups of children with the same disease and the same treatment. In those unable to stand, whether children or adults, a chair scale can be used to record weight, and recumbent anthropometric techniques can be applied (Chumlea, Roche & Mukherjee, 1984b[30], 1986[31]; Chumlea *et al.*, 1985b[36]; Chumlea, 1988[10]; Chumlea & Roche, 1984[26], 1988[28]). These techniques are preferable to a system of measurements made when such individuals are sitting in wheelchairs because the latter measurements have larger errors and many measurements must be omitted. As for anthropometry in other children, but even more so, measurements of disabled children should be repeated until the differences between pairs of values are less than set tolerances. This is particularly important in such children because the differences between paired measurements tend to be large.

### Nutritional assessment in the elderly

Much innovative work done at Fels on the nutritional assessment of the elderly has been summarized previously (Chumlea *et al.* 1985b[36]; 1990[41]; Chumlea & Roche, 1988[28]; Chumlea, Roche & Steinbaugh, 1989b[35]; Chumlea, 1991a[11], 1991b[12]). Although the data were not obtained from Fels participants, the concepts and procedures that were developed are linked closely to those used to study Fels participants. Chumlea and Baumgartner (1989[14]) reviewed published anthropometric and body composition data for the elderly and concluded that many standing and recumbent measurements could be combined in a study because the differences between them were not significant. The interobserver errors for recumbent anthropometric data in the elderly were larger than those for corresponding standing measurements of younger individuals so that recumbent measurements should be repeated to obtain equivalent levels of reliablility (Chumlea, Roche & Rogers, 1984c[32]; Chumlea, Roche & Steinbaugh, 1985a[34]). As stated by Chumlea and Baumgartner, relative weight and weight/stature$^2$ should be based on the measured present stature of an elderly person, not the stature assumed for young adulthood. To assist the assessment of nutritional status in elderly individuals who are unable to stand, Chumlea and his colleagues presented equations to predict stature from knee height, and to predict weight from a combination of anthropometric values (Chumlea *et al.*, 1985a[34]; Chumlea *et al.*, 1985c[40], 1986[31], 1988c[22]; Chumlea, 1991a[11]). Furthermore, they presented ways in which segmental impedance could be developed for this purpose (Chumlea, 1991b[12]; Chumlea, Baumgartner & Vellas, 1991[20]).

The relationships of total body fat to skinfold thicknesses seemed to be less close in the elderly than in young adults, but the relationships between abdominal circumference and total body fat appeared to be closer (Roche *et al.*, 1981b[265]). These differences between younger and older adults are important in the interpretation of anthropometric values. The increased difficulty of measuring skinfolds in the elderly is due to poor separation between the subcutaneous adipose tissue and the deep fascia and the greater compressibility of skinfolds which may underlie these differences. Skinfold thicknesses or W/S$^2$ used alone are not effective predictors of body fat in all age groups. Furthermore, $W/S^2$ may lack specificity because it is related to both body fat and fat-free mass in some age groups, for example adolescent boys.

Much work is needed to improve the assessment of nutritional status. Too often choices of procedures are made from those existing – some new procedures should be developed for use in surveys. These procedures include B-mode ultrasound and segmental impedance. Emphasis should be on validation, which has been difficult in the disabled because they cannot be weighed underwater. Data from dual photon x-ray absorptiometry should prove valuable in this regard.

### Influences on body composition
#### Familial and genetic influences

Work done at Fels relating to familial influences on body composition has been reviewed in detail by Siervogel (1983). Himes, Roche and Garn participated in a meeting of The Committee on Nutrition of the Mother and Pre-school Child in 1978 at which Garn, drawing on his Fels experience, provided heritability estimates for fatness based on intra-familial correlations. Typically, these estimates neglect environmental–genetic interactions and describe only the proportion of the variance explained by the additive component of genetic variance. Garn found low but significant correlations between mothers and infants at birth for weight and weight-for-stature and for estimates of fat-free mass from chest breadth. Interestingly, these findings were partly dependent on the length of gestation, which was slightly correlated with maternal fat-free mass ($r = 0.2$), but the fat-free mass of the father was not correlated with that of the infant at birth (Garn, 1961c[59], 1962b[61]; Garn *et al.*, 1960a[74]; Kagan & Garn, 1963). Since, in these studies, fat-free mass was estimated from chest breadth, the correlations reflected associations for chest breadth; those for fat-free mass were inferential. There were also significant sibling and spouse correlations for chest breadth (Garn, 1962a[60]).

Reynolds (1951) and Garn (1962a[60]) did not find a tendency for higher correlations between adipose tissue thicknesses at various sites in pairs of related individuals than in unrelated individuals, indicating an absence of significant genetic control over patterns of adipose distribution. Nevertheless, Reynolds (1951) showed that differences between sites in standard scores for adipose tissue thicknesses were progressively smaller as pairs of relatives sharing larger percentages of genes were considered. This topic requires more attention.

Siervogel *et al.* (1984) described serial values of $W/S^2$ and concluded that most of the family resemblance was due to shared  environments because the parent–offspring correlations were near zero but the sibling correlations were significant for $W/S^2$ at birth and 8 years, and for ages at minimum and maximum velocity of $W/S^2$. Quite correctly, however, Garn (1962a[60]) pointed out that familial analyses of $W/S^2$ are difficult to interpret because weight and stature may be independently subject to many genetic and environmental influences.

Byard and colleagues (1983, 1989) reported familial correlations that suggested sex linkage for $W/S^2$ and calf circumference but not for stature. Byard *et al.* (1988) reported age trends in the transmission of $W/S^2$ based on correlations calculated by maximum likelihood methods at annual intervals from 1 to 18 years for parent–offspring, sibling, and cousin pairs. A model in which the transmission is greater to the opposite sex than to the same sex gave the best fit before puberty. This pattern, which is consistent with X-linked inheritance, was supported by other findings that suggested the presence of a gene for greater values of $W/S^2$ that had a frequency of about 3%. This gene would affect the phenotype in 3% of boys but only 0.9% of girls. Correspondingly, there was a bimodal distribution of $W/S^2$ in boys but not girls before puberty. After puberty, an autosomic polygenic model fitted well regardless of sex, with hereditability estimates of about 40% at ages older than 10 years.

### Maternal influences
Maternal weight at 3 months of gestation and the maternal weight gain during pregnancy were not significantly correlated with the thickness of subcutaneous adipose tissue in the infant, but there were low significant correlations between these maternal influences and birth weight (Garn, 1958a[52]).

### Dietary influences
Mathematical models were used to describe the patterns of change in weight during infancy. The parameters of these models did not differ significantly between breast-fed and formula-fed infants (Himes, 1978b). This work was done in combination with a study of milk intakes and feeding patterns of breast-fed infants (Pao, Himes & Roche, 1980). It should be noted also that Garn (1967[69]) found that the caloric intakes of adults in the Fels study were about 15% lower than the US Recommended Daily Allowances. This is not surprising since these allowances are set at levels considered adequate for a large majority of the population. The lower intakes at Fels were not due to differences between the mean weight of the Fels sample and the reference weights used to develop the allowances. In

Garn's data, the caloric intakes decreased by about 5% per decade of age and were lower in women with an artificial menopause than in age-matched pre-menopausal or post-menopausal women. His data indicated that occupation was a more important influence on caloric intake than either stature or weight.

### Basal metabolic rate

It was shown that basal metabolic rate was slightly higher in males than females to 7 years, but the differences were large enough at older ages to require sex-specific reference data (Garn, Clark & Harper, 1953[73]). These sex differences were reduced after the data had been corrected for body size or surface area but differences remained that were probably due to variations in body composition. Fat-free mass, indexed by the thickness of muscle in the calf, was a major determinant of basal metabolic rate ($r = 0.8$). Nevertheless, basal metabolic rate per kilogram of body weight was higher in boys than girls matched for calf muscle thickness. These sex differences were associated with differences in the urinary excretion of ketosteroids (Clark & Garn, 1954). Basal metabolic rate was also related to skin temperature which can be measured accurately by a thermistor placed in close contact with the skin (Clark & Trolander, 1954; Garn, 1954b[43]). Furthermore, it was shown that gains in basal metabolic rates during pregnancy were related to pre-pregnancy rates; the gains were larger in those who had low pre-pregnancy rates (Sontag *et al.*, 1944[299]).

The general topic of factors that influence body composition is likely to receive much more attention at Fels in the future than in the past. Areas of special concentration are likely to include genetic influences and habitual physical activity.

### Risk factors for cardiovascular disease

Particularly since 1976, there has been considerable interest in risk factors for cardiovascular disease within the Fels Longitudinal Study. The factors that will be considered here are blood pressure and levels of plasma 'lipids.'

### Blood pressure

Familial and genetic studies related to hypertension were reviewed critically by Siervogel (1984). Roche *et al.* (1980[229]) investigated long-term changes in blood pressure using data from the major US growth studies. Systolic blood pressure became higher in boys than in girls at about 3 months, but there was an opposite sex difference from 2 to 7 years. The median values for systolic blood pressure increased considerably in boys but not girls from 14 to 18 years and there were gradual increases in both men and women. The medians for men were higher than those for women by about 12 mmHg.

Diastolic blood pressures, like systolic blood pressures, increased during infancy and childhood with the values for boys exceeding those for girls after 6 months (Roche *et al.*, 1980[229]). The sex differences in diastolic pressure were small until about 14 years, when the means became considerably larger for boys than girls; this difference persisted into adulthood. As for systolic blood pressure, the medians for girls were essentially stable after 15 years. There was significant skewness and kurtosis in systolic and diastolic blood pressures after 10 years. This was equally marked in each sex for systolic blood pressures but was more marked for females in diastolic blood pressures. During infancy, median pulse pressures (systolic pressure less diastolic pressure) changed little with age and the sex differences were inconsistent. Pulse pressures became larger for boys than girls at about 15 years and this difference continued into adulthood (Roche *et al.*, 1980[229]).

The between-subject variability of systolic and diastolic blood pressures, as measured by the standard deviations, decreased to 6 months and then slowly increased until 14 years, with little sex difference. At about 14 years, the standard deviations became larger in boys than in girls, particularly for diastolic pressure. In men, the standard deviations for systolic pressure decreased with age but those for diastolic pressure did not change, while in women, the standard deviations increased with age for both systolic and diastolic pressures.

There were low correlations between blood pressure and stature within age groups (Roche *et al.*, 1980[229]; Siervogel *et al.*, 1982c). For systolic pressure, these correlations changed from r = −0.4 at 2 years to r = 0.3 at 7 years and then remained stable in each sex until 15 years (Roche *et al.*, 1980[229]). The correlations then decreased to about r = −0.3 at 26 years after which they were stable. There were low positive correlations between diastolic pressure and stature at all ages except for men, in whom there were low negative correlations. The correlations between systolic and diastolic pressures with weight and subcutaneous adipose tissue thicknesses were low in each sex at all ages (Roche *et al.*, 1980[229]).

There were low positive correlations between systolic or diastolic blood pressure and relative skeletal age (skeletal age less chronological age) (Roche *et al.*, 1980[229]). Additionally, age at menarche and age at peak height velocity had low negative correlations with blood pressure, especially from 8 to 14 years. These findings indicated that rapidly maturing children had a slight tendency to higher blood pressures during pubescence that was more marked for systolic than for diastolic blood pressure.

There were low parent–offspring and sibling correlations for either systolic or diastolic blood pressure, measured at the same age; some of this association may be due to common effects of body size and rates of maturation, but the sibling correlations for rates of change from 9 to 18 years were not significant (Roche *et al.*, 1980[229]; Woynarowska *et al.*, 1985).

The possible tracking of blood pressure is important because of its relevance to the early institution of programs to prevent adult hypertension in those at risk. Age-to-age correlations for values 1 year apart were similar for systolic and diastolic blood pressure. These correlations increased from about 0.2 in infancy to 0.5 at 10 years but changed little after this. More importantly, correlations between childhood values and blood pressure at 18 or 30 years increased rapidly with the age of the earlier measurement until the first measurement was at the age of 4 years after which the increases were small. These correlations were higher for systolic pressure than for diastolic pressure. Finally, multiple regressions of childhood stature and weight on adult blood pressures had low $R^2$ values with negative coefficients for stature but positive coefficients for weight (Roche *et al.*, 1980[229]). This indicated a slight tendency for short heavy children to have higher blood pressures in adulthood.

The tracking of Fels blood pressure values was addressed again by Smith *et al.* (1990; in press), who fitted a model to the serial values during childhood. This work showed tracking correlations of about 0.4 for 4-year intervals and 0.2 for 20-year intervals. The relative risk of hypertension at the age of 35 years was estimated at 1.9 for 15-year-old males and 2.6 for 15-year-old females with a true diastolic pressure of 80 mmHg, in comparison with this risk of hypertension in those with a diastolic pressure of 60 mmHg.

### Relationships of blood pressure to body composition

Siervogel *et al.* (1982c) reported associations of systolic and diastolic blood pressures with body composition in children and adults. The correlations with fat-free mass and adipocyte volume were significant in children but not in adults. Furthermore, the correlations between diastolic pressure and total body fat, percent body fat or adipocyte number were not significant in adults. Baumgartner *et al.* (1987b), however, reported significant but low correlations of systolic and diastolic blood pressures with percent body fat in adults. After adjustments for age and sex, fatness appeared to be related to blood pressure, but the correlations were low.

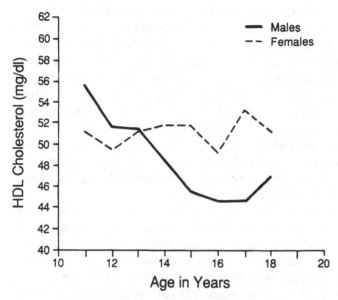

Fig. 7.12 Mean values of high density lipoprotein (HDL) cholesterol for boys and girls. (Baumgartner, R.N., Siervogel, R.M., Chumlea, W.C. & Roche, A.F. (1989c). Associations between plasma lipoprotein cholesterols, adiposity and adipose tissue distribution during adolescence. *International Journal of Obesity*, **13**, 31–41. Redrawn with permission from MacMillan Press Ltd.)

The sensitivity and specificity of the various methods used to describe adiposity and adipose tissue distributions in the context of hypertension have not been evaluated. Therefore, as noted by Siervogel and Baumgartner (1988), the reported studies may not have used the best descriptors to elucidate relationships between adiposity, adipose tissue distribution and blood pressure. Baumgartner *et al.* (1987b) used the logarithm of the ratio subscapular/lateral calf skinfold thicknesses which was correlated highly with a more complex index based on a principal components analysis of multiple skinfold thicknesses. They found that, after controlling for age, age$^2$ and percent body fat, and plasma lipids, a centripetal (truncal) pattern has low positive correlations with systolic and diastolic blood pressures in women.

### Levels of plasma 'lipids'
The means for plasma levels of high-density lipoprotein (HDL) cholesterol tended to be slightly higher in boys than girls from 11 to 13 years but not from 13 to 18 years (Baumgartner *et al.*, 1989b, 1989c). These sex differences were statistically significant only at 17 and 18 years (Fig. 7.12).

There were low correlations between plasma lipid levels and blood pressure in children after removing effects of age, fatness and body size. Girls in the lowest quartile for high-density lipoprotein had the highest systolic blood pressures and there were low but significant negative correlations between these values. The corresponding correlations for boys were not significant (Siervogel *et al.*, 1981). In another    study of adolescent children, Baumgartner *et al.* (1989c) showed that the changes with age in HDL cholesterol had significant associations with changes in adipose tissue distribution in boys but not girls. These changes in adipose tissue distribution were independent of changes in total body fat. In men and women there were associations between the degree of obesity and plasma cholesterol levels (Thomas & Garn, 1960).

An analysis of tracking from 9 to 21 years in serum levels of lipids and lipoproteins showed that the tracking coefficients for intervals of 2 to 10 years were about 0.5 to 0.7 for total cholesterol, low-density lipoprotein cholesterol and high-density lipoprotein cholesterol (Guo *et al.*, in press). This significant tracking was reflected in very strong relative risks of high values for serum lipid and lipoproteins at 21 years of age in those with high values at 9 years.

During adolescence there are changes in body composition and in the levels of circulating hormones, such as testosterone and estrogen, that directly influence the rate of maturation. Therefore, associations between lipid levels and maturity status would be expected. Nevertheless, the correlations between maturity levels and plasma lipids during adolescence were not significant within chronological age groups in either sex, but changes in lipid and lipoprotein levels were more apparent between skeletal age groups than between chronological age groups (Siervogel *et al.*, 1989b).

Fels research into body composition has been intense for the past 14 years, yet, in some ways, it is just beginning. Many data await analysis and numerous possibilities exist for the exploration of novel subject areas using new techniques, old skills, and clear thoughts.

# Epilogue

The future of the Fels Longitudinal Study is uncertain. One can look ahead with great expectations but it is 'foolish to look further than you can see,' as stated by Winston Churchill. The study is more exciting than ever: more data, more brilliant young investigators, more health-related and socially significant questions to be addressed. Consequently, the near future is viewed with optimism that is based, in part, on the knowledge that what we have done in the last few decades has contributed to knowledge, met with the approval of review groups and has received support from the Federal Government and others.

It is impossible, however, to predict how long this happy state of affairs will last. The National Institutes of Health, which provide most of the funding for the Fels study, have support cycles that do not extend longer than 5 years and funding is highly competitive. Therefore, there is some apprehension. Our view of the Fels study is biased and scientific progress becomes more difficult when all the low apples have been picked. Anything may occur. As a Swedish proverb states: 'The afternoon knows what the morning never suspected.' Despite uncertainties, we will continue as long as possible. An observational longitudinal study becomes more valuable as its duration increases and more complete descriptions of natural changes become possible, leading to more complete understanding of human development.

I wish to conclude with three quotations that are apposite. Venerable Bede wrote: 'It is better never to begin a good work than, having begun it, to stop.' A similar thought was expressed by Sir Francis Drake: 'It is not the beginning of the task, but the continuing of the same until it be well and truly finished wherein lies the true glory.' These are bold words. We do not seek glory. Rather we are mindful of our modest role and would echo the words of Isaac Newton: 'We do not know how we appear to the world, but to ourselves we seem to have been only like children playing on the seashore, and diverting ourselves in now and then finding a pebble or a prettier shell than ordinary, whilst the great ocean of truth lay all undiscovered before us.'

# References

References for the following authors have been numbered below, and have corresponding superscript numbers in the text, to enable the reader to identify them more easily: Chumlea (1–41), Garn (42–172), Roche (173–275), Sontag (276–308).

*Not based on data from the Fels Longitudinal Study.

Abdel-Malek, A. K., Mukherjee, D. & Roche, A. F. (1985). A method of constructing an index of obesity. *Human Biology*, **57**, 415–30.

Baumgartner, R. N. (1988). Associations between plasma lipoprotein cholesterols, adiposity, and adipose tissue distribution during adolescence. *American Journal of Physical Anthropology*, **75**, 184.

Baumgartner, R. N., Chumlea, W. C., Guo, S. & Roche, A. F. (1990). Prediction of growth in fat-free mass from bioelectric resistance and anthropometry in children. *American Journal of Human Biology*, **2**, 195.

Baumgartner, R. N., Chumlea, W. C. & Roche, A. F. (1987a). Associations between bioelectric impedance and anthropometric variables. *Human Biology*, **59**, 235–44.

Baumgartner, R. N., Chumlea, W. C. & Roche, A. F. (1988a). Bioelectric impedance phase angle and body composition. *American Journal of Clinical Nutrition*, **48**, 16–23.

Baumgartner, R. N., Chumlea, W. C. & Roche, A. F. (1989a). Estimation of body composition from bioelectric impedance of body segments. *American Journal of Clinical Nutrition*, **50**, 221–6.

Baumgartner, R. N., Chumlea, W. C. & Roche, A. F. (1990). Bioelectric impedance for body composition. *Exercise and Sport Sciences Reviews*, **18**, 193–224.

Baumgartner, R. N., Heymsfield, S. B., Roche, A. F. & Bernardino, M. (1988b). Abdominal composition quantified by computed tomography. *American Journal of Clinical Nutrition*, **48**, 936–45.

Baumgartner, R. N., Roche, A. F., Chumlea, W. C., Siervogel, R. M. & Glueck, C. J. (1987b). Fatness and fat patterns: Associations with plasma lipids and blood pressures in adults, 18 to 57 years of age. *American Journal of Epidemiology*, **126**, 614–28.

Baumgartner, R. N., Roche, A. F., Guo, S., Lohman, T., Boileau, R. A. & Slaughter, M. H. (1986b). Adipose tissue distribution: The stability of principal components by sex, ethnicity and maturation stage. *Human Biology*, **58**, 719–35.

Baumgartner, R. N., Roche, A. F. & Himes, J. H. (1986a). Incremental growth tables. *American Journal of Clinical Nutrition*, **43**, 711–22.

Baumgartner, R. N., Siervogel, R. M., Chumlea, W. C. & Roche, A. F. (1989c).

Associations between plasma lipoprotein cholesterols, adiposity and adipose tissue distribution during adolescence. *International Journal of Obesity*, **13**, 31–41.

Baumgartner, R. N., Siervogel, R. M. & Roche, A. F. (1989b). Clustering of cardiovascular risk factors in association with indices of adiposity and adipose tissue distribution in adults. *American Journal of Human Biology*, **1**, 43–62.

Bernard, J. & Sontag, L. W. (1947). Fetal reactivity to tonal stimulation: A preliminary report. *Journal of Genetic Psychology*, **70**, 205–10.

*Björck, A. (1967). Solving linear least squares problems by Gran–Schmidt orthogonalization. *Nordisk Tijdschrift Informationsbehandlung*, **7**, 1–21.

Bock, R. D. (1982). Predicting the mature stature of preadolescent children. In *Genetic and Environmental Factors During the Growth Period*, ed. C. Susanne, pp. 3–19. New York: Plenum Press.

Bock, R. D. (1986). Unusual growth patterns in the Fels data. In *Human Growth: A Multidisciplinary Review*, ed. A. Demirjian, pp. 69–84. London: Taylor & Francis.

Bock, R. D. & Sykes, R. C. (1989). Evidence for continuing secular increase in height within families in the United States. *American Journal of Human Biology*, **1**, 143–8.

Bock, R. D., Sykes, R. C. & Roche, A. F. (1985). *The Fels Three-generation data: evidence for continuing secular increase in height.* Abstract No. 11.22, p. 63, IVth International Congress of Auxology, Montreal, Quebec, Canada, June 16–20.

*Bock, R. D. & Thissen, D. (1976). Fitting multi-component models for growth in stature. *Proceedings of the 9th International Biometric Conference*, **1**, 432–42.

Bock, R. D. & Thissen, D. (1980). Statistical problems of fitting individual growth curves. In *Human Physical Growth and Maturation: Methodologies and Factors*, ed. F. E. Johnston, A. F. Roche & C. Susanne, pp. 268–90. New York: Plenum Press.

Bock, R. D., Wainer, H. C., Peterson, A., Thissen, D., Murray, J. & Roche, A. F. (1973). A parametrization for individual human growth curves. *Human Biology*, **45**, 63–80.

Boothby, W. M., Berkson, J. & Dunn, H. L. (1936). Studies of the energy of metabolism of normal individuals; a standard for basal metabolism, with a nomogram for clinical application. *American Journal of Physiology*, **116**, 468–84.

*Box, G. E. P. & Cox, D. R. (1964). An analysis of transformations. *Journal of the Royal Statistical Society*, Series B 26, 211–52.

*Broadbent, B. H., Sr. (1931). A new X-ray technique and its application to orthodontia. *Angle Orthodontist*, **1**, 45–66.

*Broadbent, B. H., Sr., Broadbent, B. H., Jr. & Golden, W. H. (1975). *Bolton Standards of Dentofacial Developmental Growth.* St Louis, Missouri: C. V. Mosby.

Byard, P. J., Guo, S. & Roche, A. F. (1991). Model fitting to early childhood length and weight data from the Fels Longitudinal Study of Growth. *American Journal of Human Biology*, **3**, 33–40.

Byard, P. J., Ohtsuki, F., Siervogel, R. M. & Roche, A. F. (1984). Sibling correlations for serial cranial measurements from radiographs. *Journal of Craniofacial Genetics and Developmental Biology*, **4**, 265–9.

Byard, P. J. & Roche, A. F. (1984). Secular trend for recumbent length and stature in the Fels Longitudinal Growth Study. In *Human Growth and Development*, ed. J. Borms, R. Hauspie, A. Sand, C. Susanne & M. Hebbelinck, pp. 209–14. New York: Plenum Press.

Byard, P. J., Siervogel, R. M. & Roche, A. F. (1983). Familial correlations for serial measurements of recumbent length and stature. *Annals of Human Biology*, **10**, 281–93.

Byard, P. J., Siervogel, R. M. & Roche, A. F. (1988). Age trends in transmissible and nontransmissible components of family resemblance for stature. *Annals of Human Biology*, **15**, 111–18.

Byard, P. J., Siervogel, R. M. & Roche, A. F. (1989). X-linked pattern of inheritance of serial measures of weight/stature$^2$. *American Journal of Human Biology*, **1**, 443–49.

1. Chumlea, W. C. (1981). Associations between changes in adipose cellularity and total body fatness in adolescence. *American Journal of Physical Anthropology*, **54**, 208–9.

2. Chumlea, W. C. (1982). Physical growth in adolescence. In *Handbook of Developmental Psychology*, ed. B. B. Wolman, G. Stricker, S. J. Ellman, P. Keith-Spiegel & D. S. Palermo, pp. 471–85. Englewood Cliffs, New Jersey: Prentice Hall.

3. Chumlea, W. C. (1983). *Development of pelvic bone shape and tissue in children*. DTNH22-83-P-07365. US Department of Transportation, Washington, DC: US Government Printing Office.

4. Chumlea, W. C. (1984). Growth of the pelvis in children. *Transactions of the Society of Automotive Engineers*, **4**, 649–59.

5. Chumlea, W. C. (1985a). Accuracy and reliability of a new sliding caliper. *American Journal of Physical Anthropology*, **68**, 425–7.

6. Chumlea, W. C. (1985b). Assessment of body composition in nonambulatory persons. In *Body Composition Assessments in Youth and Adults*, Report of the Sixth Ross Conference on Medical Research, ed. A. F. Roche, pp. 86–90. Columbus, Ohio: Ross Laboratories.

7. Chumlea, W. C. (1985c). New methods of nutritional assessment of the elderly. *Consultant Dietician Newsletter*, **10**, 19–22.

8. Chumlea, W. C. (1986a). Assessing growth and nutritional status in children with developmental disabilities. *Consultant Dietician Newsletter*, **11**, 23–5.

9. Chumlea, W. C. (1986b). Clinical methods of assessing obesity in children. *Perspectives in Lipid Disorders*, **3**, 14–19.

10. Chumlea, W. C. (1988). Methods of nutritional anthropometric assessment for special groups. In *Anthropometric Standardization Reference Manual*, ed. T. Lohman, A. F. Roche & R. Martorell, pp. 93–5. Champaign, Illinois: Human Kinetic Books.

11. Chumlea, W. C. (1991a). Anthropometric assessment of nutritional status in the elderly. In *Anthropometric Assessment of Nutritional Status*, ed. J. Himes, pp. 399–418. New York: Wiley-Liss.

12. Chumlea, W. C. (1991b). Bioelectric impedance in the elderly. Possibilities and pitfalls. *Age and Nutrition*, **2**, 4–6.

13. Chumlea, W. C. (in press). Growth and development. In *Pediatric Nutrition: A Reference Manual*, ed. C. E. Lang & P. Queen. New York: Aspen Publishers.
14. Chumlea, W. C. & Baumgartner, R. N. (1989). Status of anthropometry and body composition data in elderly subjects. *American Journal of Clinical Nutrition*, **50**, 1158–66.
15. Chumlea, W. C. & Baumgartner, R. N. (1990). Bioelectric impedance methods for the estimation of body composition. *Canadian Journal of Sports Science*, **15**, 172–9.
16. Chumlea, W. C., Baumgartner, R. M. & Roche, A. F. (1986). Segmental bioelectric impedance measures of body composition. Poster, International Conference on 'In Vivo Body Composition Studies,' Brookhaven, New York.
17. Chumlea, W. C., Baumgartner, R. N. & Roche, A. F. (1987b). Segmental bioelectric impedance measures of body composition. In *In Vivo Body Composition Studies*, ed. K. J. Ellis, S. Yasumura & W. D. Morgan, pp. 103–107. Proceedings of an International Symposium held at Brookhaven National Laboratory, New York. *Institute of Physicists in Science and Medicine*, **47**, 35.
18. Chumlea, W. C., Baumgartner, R. N. & Roche, A. F. (1988a). Specific resistivity used to estimate fat-free mass from segmental body measures of bioelectric impedance. *American Journal of Clinical Nutrition*, **48**, 7–15.
19. Chumlea, W. C., Baumgartner, R. N. & Roche, A. F. (1988b). Fat volumes and total body fat from impedance. *Medicine and Science in Sports and Exercise*, **20**, S82.
20. Chumlea, W. C., Baumgartner, R. N. & Vellas, B. P. (1991). Anthropometry and body composition in the perspective of nutritional status in the elderly. *Nutrition*, **7**, 57–60.
21. Chumlea, W. C., Falls, R. A. & Webb, P. (1982a). Body composition in adults after 60 years of age. *American Journal of Physical Anthropology*, **57**, 176–7.
22. Chumlea, W. C., Guo, S., Roche, A. F. & Steinbaugh, M. L. (1988c). Prediction of body weight for the nonambulatory elderly from anthropometry. *Journal of the American Dietetic Association*, **88**, 564–8.
23. Chumlea, W. C. & Knittle, J. L. (1980). Associations between fat cellularity and total body fatness in adolescents and adults. *American Journal of Physical Anthropology*, **52**, 213.
24. Chumlea, W. C., Knittle, J. L., Roche, A. F., Siervogel, R. M. & Webb, P. (1981a). Size and number of adipocytes and measures of body fat in boys and girls 10 to 18 years of age. *American Journal of Clinical Nutrition*, **34**, 1791–7.
25. Chumlea, W. C., Mukherjee, D. & Roche, A. F. (1984a). A comparison of methods for measuring cortical bone thickness. *American Journal of Physical Anthropology*, **65**, 83–6.
26. Chumlea, W. C. & Roche, A. F. (1984). Nutritional anthropometric assessment of non-ambulatory persons using recumbent techniques. *American Journal of Physical Anthropology*, **63**, 146.
27. Chumlea, W. C. & Roche, A. F. (1986). Ultrasonic and skinfold caliper measures of subcutaneous adipose tissue thickness in elderly men and women. *American Journal of Physical Anthropology*, **71**, 351–7.
28. Chumlea, W. C. & Roche, A. F. (1988). Assessment of the nutritional status of

healthy and handicapped adults. In *Anthopometric Standardization Reference Manual*, ed. T. Lohman, A. F. Roche & R. Martorell, pp. 115–19. Champaign, Illinois: Human Kinetics Books.

29. Chumlea, W. C., Roche, A. F., Guo, S. & Woynarowska, B. (1987a). The influence of physiological variables and oral contraceptives on bioelectric impedance. *Human Biology*, **59**, 257–69.

30. Chumlea, W. C., Roche, A. F. & Mukherjee, D. (1984b). *Nutritional Assessment of the Elderly through Anthropometry*. Columbus, Ohio: Ross Laboratories.

31. Chumlea, W. C., Roche, A. F. & Mukherjee, D. (1986). Some anthropometric indices of body composition for elderly adults. *Journal of Gerontology*, **41**, 36–9.

32. Chumlea, W. C., Roche, A. F. & Rogers, E. (1984c). Replicability for anthropometry in the elderly. *Human Biology*, **56**, 329–37.

33. Chumlea, W. C., Roche, A. F., Siervogel, R. M., Knittle, J. L. & Webb, P. (1981b). Adipocytes and adiposity in adults. *American Journal of Clinical Nutrition*, **34**, 1798–803.

34. Chumlea, W. C., Roche, A. F. & Steinbaugh, M. L. (1985a). Estimating stature from knee height for persons 60 to 90 years of age. *Journal of the American Geriatric Society*, **33**, 116–20.

35. Chumlea, W. C., Roche, A. F. & Steinbaugh, M. L. (1989b). Anthropometric approaches to the nutritional assessment of the elderly. In *Human Nutrition and Aging*, ed. H. N. Munro, pp. 335–61. New York: Plenum Press.

36. Chumlea, W. C., Roche, A. F., Steinbaugh, M. L. & Gopalaswamy, N. (1985b). Nutritional assessment of the elderly by recumbent anthropometric methods. In *Nutrition, Immunity and Illness in the Elderly*, ed. R. K. Chandra, pp. 53–61. New York: Pergamon Press.

37. Chumlea, W. C., Roche, A. F. & Thissen, D. (1989a). The FELS method of assessing the skeletal maturity of the hand–wrist. *American Journal of Human Biology*, **1**, 175–83.

38. Chumlea, W. C., Siervogel, R. M., Roche, A. F., Mukherjee, D. & Webb, P. (1982b). Changes in adipocyte cellularity in children ten to 18 years of age. *International Journal of Obesity*, **6**, 383–9.

39. Chumlea, W. C., Siervogel, R. M., Roche, A. F., Webb, P. & Rogers, E. (1983). Increments across age in body composition of children 10 to 18 years of age. *Human Biology*, **55**, 845–52.

40. Chumlea, W. C., Steinbaugh, M. L., Roche, A. F., Mukherjee, D. & Gopalaswamy, N. (1985c). Nutritional anthropometric assessment in elderly persons 65 to 90 years of age. *Journal of Nutrition in the Elderly*, **4**, 39–51.

41. Chumlea, W. C., Vellas, B. J., Roche, A. F., Guo, S. & Steinbaugh, M. (1990). Paticularités et intéret des measures anthropométriques du status nutritionnel des personnes agées. *Age and Nutrition*, **1**, 7–12.

Clark, L. C., Jr. & Beck, E. (1950). Plasma 'alkaline' phosphatase activity. I. Normative data for growing children. *Journal of Pediatrics*, **36**, 335–41.

Clark, L. C., Jr., Beck, E. I. & Shock, N. W. (1951a). Serum alkaline phosphatase in middle and old age. *Journal of Gerontology*, **6**, 7–12.

Clark, L. C., Jr., Beck, E. & Thompson, H. (1951b). The excretion of acid phosphatase as a measure of prostatic development during pubescence. *Journal of Clinical Endocrinology*, **11**, 84–90.

Clark, L. C., Jr. & Garn, S. M. (1954). Relationship between ketosteroid excretion and basal oxygen consumption in children. *Journal of Applied Physiology*, **6**, 546–50.

Clark, L. C., Jr. & Trolander, H. (1954). Thermometer for measuring body temperature in hypothermia. *Journal of the American Medical Association*, **155**, 251–2.

Cochram, D. B. & Baumgartner, R. N. (1990). Evaluation of accuracy and reliability of calipers for measuring recumbent knee height in elderly people. *American Journal of Clinical Nutrition*, **52**, 397–400.

Colbert, C. (1968). *Bone Mineral Measurements from X-ray Images*, NIH Conference on Progress in Methods of Bone Mineral Measurements, ed. G. D. Whedon, pp. 80–95, Bethesda, Maryland.

Colbert, C. (1974). Radiographic absorptiometry. PhD Thesis, University Without Walls, Antioch College, Yellow Springs, Ohio.

Colbert, C., Bachtell, R. S., Bailey, J., Spencer, R. & Himes, J. H. (1980). Comparison of radiogrammetry with 'Radiographic Absorptiometry' (photo-densitometry). In *Proceedings of the Fourth International Conference on Bone Measurement*, pp. 439–40. *Washington, DC: US Department of Health and Human Services, US Government Printing Office.*

Colbert, C. & Garrett, C. (1969). Photodensiometry of bone roentgenograms with an on-line computer. *Journal of Clinical Orthopedics*, **65**, 39–45.

Colbert, C., Israel, H. & Garn, S. M. (1966). Absolute radiographic densities of dentin and enamel. *Journal of Dental Research*, **45**, 1826.

Colbert, C., Mazess, R. B. & Schmidt, P. B. (1970b). Bone mineral determination in vitro by radiographic photodensitometry and direct photon absorptiometry. *Investigative Radiology*, **5**, 336–40.

Colbert, C., Schmidt, P. B. & Mazess, R. B. (1969). Bone mineral determination: A comparison of radiographic photodensitometry and direct photon absorptiometry. *American Association of Physicists in Medicine, Quarterly Bulletin*, **3**, 25.

Colbert, C., Spruit, J. J. & Davila, L. R. (1967). Biophysical properties of bone; determining mineral concentrations from the X-ray image. *New York Academy of Science, Transactions*, **30**, 271–90.

Colbert, C., Spruit, J. J. & Davila, L. R. (1970a). Fels microdensitometer/ computer for bone mineral determination from roentgenograms. *Critical Reviews in Radiological Sciences*, **1**, 459–71.

Colbert, C., Van Hulst, H. & Spruit, J. J. (1970c). A phalangeal atlas: An application of radiographic photodensiometry. In *Proceedings of the Bone Mineral Conference*, ed. J. R. Cameron, pp. 224–35. Madison, Wisconsin: University of Wisconsin.

Coleman, W. H. (1969). Sex differences in the growth of the human bony pelvis. *American Journal of Physical Anthropology*, **31**, 125–52.

Crandall, V. C. (1972). The Fels Study: Some contributions to personality development and achievement in childhood and adulthood. *Seminars in Psychiatry*, **4**, 383–97.

Cronk, C. E. (1983). Fetal growth as measured by ultrasound. *Yearbook of Physical Anthropology*, **26**, 65–90.

Cronk, C. E., Mukherjee, D. & Roche, A. F. (1983b). Changes in triceps and subscapular skinfold thickness during adolescence. *Human Biology*, **55**, 707–21.

Cronk, C. E. & Roche, A. F. (1982). Race- and sex-specific reference data for triceps and subscapular skinfolds and weight/stature². *American Journal of Clinical Nutrition*, **35**, 347–54.

Cronk, C. E., Roche, A. F., Chumlea, W. C., Kent, R. & Berkey, C. (1982a). Longitudinal trends of weight/stature² in childhood in relation to adulthood body fat measures. *Human Biology*, **54**, 751–64.

Cronk, C. E., Roche, A. F., Kent, R., Berkey, C., Reed, R. B., Valadian, I., Eichorn, D. & McCammon, R. (1982b). Longitudinal trends and continuity in weight/stature² from 3 months to 18 years. *Human Biology*, **54**, 729–49.

Cronk, C. E., Roche, A. F., Kent, R., Jr., Eichorn, D. & McCammon, R. M. (1983a). Longitudinal trends in subcutaneous fat thickness during adolescence. *American Journal of Physical Anthropology*, **61**, 197–204.

*Edwards, D. A. W., Hammond, W. H., Healy, M. J. R., Tanner, J. M. & Whitehouse, R. H. (1955). Design and accuracy of calipers for measuring subcutaneous tissue thickness. *British Journal of Nutrition*, **9**, 133–43.

Falk, C., Rubinstein, P., Roche, A. F., Siervogel, R. M., Molthan, L., Fotino, M., Martin, M. & Allen, F. H. (1982). Lod scores for linkage analysis of 28 genetic markers: a survey report on 600 families. *Cytogenetics and Cell Genetics*, **32**, 272.

Falkner, F. (1971a). Skeletal maturity indicators in infancy. *American Journal of Physical Anthropology*, **35**, 393–4.

Falkner, F. (1971b). Physical growth of the child. *Carnets de l'enfance, Assignment Children, United Nations Children's Fund*, **15**, 15–22.

Falkner, F. (1972). The creation of growth standards: A committee report. *American Journal of Clinical Nutrition*, **25**, 218–20.

Falkner, F. (1973a). Velocity growth. *Pediatrics*, **51**, 746–7.

Falkner, F. (1973b). Long term developmental studies: A critique. *Early Development*, **51**, 412–21.

Falkner, F. (1975). Nutrition and bone development. *Modern Problems in Paediatrics*, **14**, 185–8.

Falkner, F. (1977a). Health Needs of Adolescents. *World Health Organization Technical Report Series*, **609**, 1–12.

Falkner, F. (1977b). Normal growth and development: Current concepts. *Postgraduate Medicine*, **62**, 58–63.

Falkner, F. (1978a). Early postnatal growth evaluation in full-term, pre-term, and small-for-dates infants. In *Auxology–Human Growth in Health and Disorder*, eds. L. Gedda & P. Parisi, pp. 79–86. New York: Academic Press.

Falkner, F. (1978b). Postnatal growth. In *Perinatal Physiology*, ed. U. Stave, pp. 37–45. New York: Plenum Press.

Falkner, F. & Roche, A. F. (1987). Relationship of femoral length to recumbent length and stature in fetal, neonatal and early childhood growth. *Human Biology*, **59**, 769–73.

Fels, S. S. (1933). *This Changing World – As I See Its Change and Purpose*. Boston: Houghton Mifflin.

Frisancho, A. R., Garn, S. M. & Ascoli, W. (1970). Subperiosteal and endosteal bone apposition during adolscence. *Human Biology*, **42**, 639–64.

42. Garn, S. M. (1954a). Fat patterning and fat intercorrelations in the adult male. *Human Biology*, **26**, 59–69.

43. Garn, S. M. (1954b). The measurement of skin temperature. *American Journal of Physical Anthropology*, **12**, 127–30.

44. Garn, S. M. (1955a). Applications of pattern analysis to anthropometric data. *New York Academy of Science Annals*, **63**, 537–52.

45. Garn, S. M. (1955b). Relative fat patterning: An individual characteristic. *Human Biology*, **27**, 75–89.

46. Garn, S. M. (1956a). Comparison of pinch-caliper and X-ray measurements of skin plus subcutaneous fat. *Science*, **124**, 178–9.

47. Garn, S. M. (1956b). Fat thickness and growth progress during infancy. *Human Biology*, **28**, 232–50.

48. Garn, S. M. (1957a). Selection of body sites for fat measurement. *Science*, **125**, 550–1.

49. Garn, S. M. (1957b). An improved method of estimating body fat content by the use of teleoroentgenograms. ASTIA Document No. AD 115 085.

50. Garn, S. M. (1957c). Fat weight and fat placement in the female. *Science*, **125**, 1091–2.

51. Garn, S. M. (1957d). Roentgenogrammetric determinations of body composition. *Human Biology*, **29**, 337–53.

52. Garn, S. M. (1958a). Fat, body size and growth in the newborn. *Human Biology*, **30**, 265–80.

53. Garn, S. M. (1958b). Statistics: A review. *Angle Orthodontist*, **28**, 149–65.

54. Garn, S. M. (1960a). The number of hand–wrist centers. *American Journal of Physical Anthropology*, **18**, 293–9.

55. Garn, S. M. (1960b). Growth and development. *50th Anniversary White House Conference on Children and Youth*, Vol. 2: *Development and Education of the Nation's Children*, ed. E. Ginzberg, pp. 24–42. New York: Columbia University Press.

56. Garn, S. M. (1960c). Fat accumulation and aging in males and females. In *The Biology of Aging*, ed. B. L. Strehler, pp. 170–80. Symposium No. 6, Gatlinburg, Tennessee, May 1–3, 1957. Washington , DC: American Institute of Biological Sciences.

57. Garn, S. M. (1961a). Research and malocclusion. *American Journal of Orthodontics*, **47**, 661–73.

58. Garn, S. M. (1961b). Radiographic analysis of body composition. In *Techniques for Measuring Body Composition*, ed. J. Brozek & A. Henschel, pp. 36–58. Washington, DC: National Academy of Sciences.

59. Garn, S. M. (1961c). The genetics of normal human growth. In *De Genetica Medica*, ed. L. Gedda, pp. 413–32. Instituto Gregorio Mendel: Rome, Italy. Reprinted in *Physical Anthropology 1953–1961*, ed. G. W. Lasker, pp. 291–310. Mexico City: Instituto de Investigaciones Historicas, Universidad Nacional Autonoma de Mexcio and Instituto Nacional e Anthropologia e Historia.

60. Garn, S. M. (1962a). Anthropometry in clinical appraisal of nutritional status. *American Journal of Clinical Nutrition*, **11**, 418–32.
61. Garn, S. M. (1962b). Determinants of size and growth during the first three years. *Modern Problems in Paediatrics*, **7**, 50–4.
62. Garn, S. M. (1962c). Automation in anthropometry. *American Journal of Physical Anthropology*, **20**, 387–8.
63. Garn, S. M. (1963a). Some pitfalls in the quantification of body composition. *New York Academy of Science Annals*, **110**, 171–4.
64. Garn, S. M. (1963b). Human biology and research in body composition. *New York Academy of Science Annals*, **110**, 429–46.
65. Garn, S. M. (1965). The applicability of North American growth standards in developing countries. *Canadian Medical Association Journal*, **93**, 914–19.
66. Garn, S. M. (1966a). The evolutionary and genetic control of variability in man. *Annals of the New York Academy of Sciences*, **134**, 602–15.
67. Garn, S. M. (1966b). Malnutrition and skeletal development in the pre-school child. In *Pre-School Malnutrition*, pp. 43–62. Washington, DC: National Academy of Sciences – National Research Council.
68. Garn, S. M. (1966c). Body size and its implications. *Review of Child Development Research*, **2**, 529–61.
69. Garn, S. M. (1967). Food intakes of older people: Recommended vs. observed. National Academy of Science – National Research Council. *Proceedings, Food and Nutrition Board*, **27**, 9–11.
70. Garn, S. M. (1970). The earlier gain and the latter loss of cortical bone. In *Nutritional Perspective*, p. 146. Springfield, Illinois: Charles C. Thomas.
71. Garn, S. M., Blumenthal, T. & Rohmann, C. G. (1965g). On skewness in the ossification centers of the elbow. *American Journal of Physical Anthropology*, **23**, 303–4.
72. Garn, S. M. & Clark, L. C., Jr. (1955). Creatinine–weight coefficient as a measurement of obesity. *Journal of Applied Physiology*, **8**, 135–8.
73. Garn, S. M., Clark, L. C., Jr. & Harper, R. V. (1953). The sex difference in the basal metabolic rate. *Child Development*, **24**, 215–24.
74. Garn, S. M., Clark, A., Landkof, L. & Newell, L. (1960a). Parental body build and developmental progress in the offspring. *Science*, **132**, 1555–6.
75. Garn, S. M., Dahlberg, A. A. & Kerewsky, R. S. (1966f). Groove pattern, cusp number and tooth size. *Journal of Dental Research*, **45**, 970.
76. Garn, S. M., Lewis, A. B., Kerewsky, R. S. & Dahlberg, A. A. (1966g). Genetic independence of Carabelli's trait from tooth size or crown morphology. *Archives of Oral Biology*, **11**, 745–7.
77. Garn, S. M., Dahlberg, A. A., Lewis, A. B. & Kerewsky, R. S. (1966h). Cusp number, occlusal groove pattern and human taxonomy. *Nature*, **210**, 224–5.
78. Garn, S. M., Davila, G. H. & Rohmann, C. G. (1968f). Population frequencies and altered remodeling mechanisms in normal medullary stenosis. *American Journal of Physical Anthropology*, **29**, 425–8.
79. Garn, S. M., Fels, S. L. & Israel, H. (1967h). Brachymesophalangia of digit five in ten populations. *American Journal of Physical Anthropology*, **27**, 205–10.
80. Garn, S. M., Feutz, E., Colbert, C. & Wagner, B. (1966c). Comparison of cortical thickness and radiographic microdensitometry in the measurement of

bone loss. In *Progress in Development of Methods in Bone Densitometry*, pp. 65–77, NASA SP-64. Washington, DC: National Aeronautics and Space Administration.

81. Garn, S. M. & French, N. Y. (1963). Postpartum and age changes in areolar pigmentation. *American Journal of Obstetrics and Gynecology*, **85**, 873–5.
82. Garn, S. M. & French, N. Y. (1967). Magnitude of secular trend in the Fels population: stature and weight. Private printing.
83. Garn, S. M. & Gorman, E. L. (1956). Comparison of pinch-caliper and teleroentgenogrammetric measurements of subcutaneous fat. *Human Biology*, **28**, 407–13.
84. Garn, S. M., Greaney, G. & Young, R. (1956d). Fat thickness and growth progress during infancy. *Human Biology*, **28**, 232–50.
85. Garn, S. M. & Harper, R. V. (1955). Fat accumulation and weight gain in the adult male. *Human Biology*, **27**, 39–49.
86. Garn, S. M. & Haskell, J. A. (1959a). Fat and growth during childhood. *Science*, **130**, 1711.
87. Garn, S. M. & Haskell, J. A. (1959b). Fat changes during adolescence. *Science*, **129**, 1615–16.
88. Garn, S. M. & Haskell, J. A. (1960). Fat thickness and developmental status in childhood and adolescence. *American Journal of Diseases of Children*, **99**, 746–51.
89. Garn, S. M. & Helmrich, R. H. (1967). Next step in automated anthropometry. *American Journal of Physical Anthropology*, **26**, 97–100.
90. Garn, S. M., Helmrich, R. H., Flaherty, K. M. & Silverman, R. N. (1967a). Skin dosages in radiation-sparing techniques for the laboratory and field. *American Journal of Physical Anthropology*, **26**, 101–6.
91. Garn, S. M., Helmrich, R. H. & Lewis, A. B. (1967i). Transducer caliper with readout capability for odontometry. *Journal of Dental Research*, **46**, 306.
92. Garn, S. M., Hempy, H. O., III & Schwager, P. M. (1968b). Measurement of localized bone growth employing natural markers. *American Journal of Physical Anthropology*, **28**, 105–8.
93. Garn, S. M., Hertzog, K. P., Poznanski, A. K. & Nagy, J. M. (1972). Metacarpophalangeal length in the evaluation of skeletal malformations. *Radiology*, **105**, 375–81.
94. Garn, S. M. & Hull, E. I. (1966). Taller individuals lose less bone as they grow older. *Investigative Radiology*, **1**, 255–6.
95. Garn, S. M., Kerewsky, R. S. & Lewis, A. B. (1966j). Extent of sex influence on Carabelli's polymorphism. *Journal of Dental Research*, **45**, 1823.
96. Garn, S. M. & Koski, K. (1957). Tooth eruption sequence in fossil and recent man. *Nature*, **180**, 442–3.
97. Garn, S. M., Koski, K. & Lewis, A. B. (1957a). Problems in determining the tooth eruption sequence in fossil and modern man. *American Journal of Physical Anthropology*, **15**, 313–31.
98. Garn, S. M. & Lewis, A. B. (1957). Relationship between the sequence of calcification and the sequence of eruption of the mandibular molar and premolar teeth. *Journal of Dental Research*, **36**, 992–5.
99. Garn, S. M. & Lewis, A. B. (1958). Tooth-size, body-size and 'giant' fossil. *American Anthropologist*, **60**, 874–80.

100. Garn, S. M. & Lewis, A. B. (1962). The relationship between third molar agenesis and reduction in tooth number. *Angle Orthodontist*, **32**, 14–18.
101. Garn, S. M. & Lewis, A. B. (1970). The gradient and the pattern of crown-size reduction in simple hypodontia. *Angle Orthodontist*, **40**, 51–8.
102. Garn, S. M., Lewis, A. B. & Bonné, B. (1961a). Third molar polymorphism and the timing of tooth formation. *Nature*, **192**, 989.
103. Garn, S. M., Lewis, A. B. & Bonné, B. (1962a). Third molar formation and its developmental course. *Angle Orthodontist*, **32**, 270–9.
104. Garn, S. M., Lewis, A. B. & Kerewsky, R. S. (1963b). Third molar agenesis and size reduction of the remaining teeth. *Nature*, **200**, 488–9.
105. Garn, S. M., Lewis, A. B. & Kerewsky, R. S. (1964a). Third molar agenesis and variation in size of the remaining teeth. *Nature*, **201**, 839.
106. Garn, S. M., Lewis, A. B. & Kerewsky, R. S. (1964b). Sex difference in tooth size. *Journal of Dental Research*, **43**, 1039.
107. Garn, S. M., Lewis, A. B. & Kerewsky, R. S. (1965a). X-linked inheritance of tooth size. *Journal of Dental Research*, **44**, 439–41.
108. Garn, S. M., Lewis, A. B. & Kerewsky, R. S. (1965b). Genetic, nutritional, and maturational correlates of dental development. *Journal of Dental Research*, **44**, 228–42.
109. Garn, S. M., Lewis, A. B. & Kerewsky, R. S. (1965f). Size interrelationships of the mesial and distal teeth. *Journal of Dental Research*, **44**, 350–54.
110. Garn, S. M., Lewis, A. B. & Kerewsky, R. S. (1966e). The meaning of bilateral asymmetry in the permanent dentition. *Angle Orthodontist*, **36**, 55–62.
111. Garn, S. M., Lewis, A. B. & Kerewsky, R. S. (1966i). Bilateral asymmetry and concordance in cusp number and crown morphology of the mandibular first molar. *Journal of Dental Research*, **45**, 1820.
112. Garn, S. M., Lewis, A. B. & Kerewsky, R. S. (1967b). Sex difference in tooth shape. *Journal of Dental Research*, **46**, 1470.
113. Garn, S. M., Lewis, A. B. & Kerewsky, R. S. (1967c). The relationship between sexual dimorphism in tooth size and body size as studied within families. *Archives of Oral Biology*, **12**, 299–302.
114. Garn, S. M., Lewis, A. B. & Kerewsky, R. S. (1967d). Buccolingual size asymmetry and its developmental meaning. *Angle Orthodontist*, **37**, 186–93.
115. Garn, S. M., Lewis, A. B. & Kerewsky, R. S. (1967j). Shape similarities throughout the dentition. *Journal of Dental Research*, **46**, 1481.
116. Garn, S. M., Lewis, A. B. & Kerewsky, R. S. (1968d). The magnitude and implications of the relationship between tooth size and body size. *Archives of Oral Biology*, **13**, 129–31.
117. Garn, S. M., Lewis, A. B. & Kerewsky, R. S. (1968g). Relationship between the buccolingual and mesiodistal tooth diameters. *Journal of Dental Research*, **47**, 495.
118. Garn, S. M., Lewis, A. B., Kerewsky, R. S. & Jegart, K. (1965c). Sex differences in intraindividual tooth size commualities. *Journal of Dental Research*, **44**, 476–9.
119. Garn, S. M., Lewis, A. B., Koski, K. & Polacheck, D. L. (1958a). The sex difference in tooth calcification. *Journal of Dental Research*, **37**, 561–7.

120. Garn, S. M., Lewis, A. B. & Polacheck, D. L. (1958b). Variability of tooth formation in man. *Science*, **128**, 1510.
121. Garn, S. M., Lewis, A. B. & Polacheck, D. L. (1959). Variability of tooth formation. *Journal of Dental Research*, **38**, 135–48.
122. Garn, S. M., Lewis, A. B. & Polacheck, D. L. (1960b). Interrelations in dental development. I. Interrelationships within the dentition. *Journal of Dental Research*, **39**, 1049–55.
123. Garn, S. M., Lewis, A. B. & Polacheck, D. L. (1960c). Sibling similarities in dental development. *Journal of Dental Research*, **39**, 170–5.
124. Garn, S. M., Lewis, A. B. & Shoemaker, D. W. (1956a). The sequence of calcification of the mandibular molar and premolar teeth. *Journal of Dental Research*, **35**, 555–61.
125. Garn, S. M., Lewis, A. B., Swindler, D. R. & Kerewsky, R. S. (1967k). Genetic control of sexual dimorphism in tooth size. *Journal of Dental Research*, **46**, 963–72.
126. Garn, S. M., Lewis, A. B. & Vicinus, J. H. (1962b). Third molar agenesis and reduction in the number of other teeth. *Journal of Dental Research*, **41**, 717.
127. Garn, S. M., Lewis, A. B. & Vicinus, J. H. (1963d). Third molar polymorphism and its significance to dental genetics. *Journal of Dental Research*, **42**, 1344–63.
128. Garn, S. M., Lewis, A. B. & Walenga, A. J. (1968a). Maximum-confidence values for the human mesiodistal crown dimension of human teeth. *Archives of Oral Biology*, **13**, 841–4.
129. Garn, S. M., Lewis, A. B. & Walenga, A. J. (1968c). Evidence for a secular trend in tooth size over two generations. *Journal of Dental Research*, **47**, 503.
130. Garn, S. M., Lewis, A. B. & Walenga, A. J. (1968e). The genetic basis of the crown-size profile pattern. *Journal of Dental Research*, **47**, 1190.
131. Garn, S. M. & Nolan, P., Jr. (1963). A tank to measure body volume by water displacement (BoVoTa). *New York Academy of Science Annals*, **110**, 91–5.
132. Garn, S. M., Pao, E. M. & Rohmann, C. G. (1965d). Calcium intake and compact bone loss in adult subjects. *Federation Proceedings*, **24**, 567.
133. Garn, S. M. & Rohmann, C. G. (1959). Communalities of the ossification centers of the hand and wrist. *American Journal of Physical Anthropology*, **17**, 319–23.
134. Garn, S. M. & Rohmann, C. G. (1960a). The number of hand–wrist centers. *American Journal of Physical Anthropology*, **18**, 293–9.
135. Garn, S. M. & Rohmann, C. G. (1960b). Variability in the order of ossification of the bony centers of the hand and wrist. *American Journal of Physical Anthropology*, **18**, 219–28.
136. Garn, S. M. & Rohmann, C. G. (1962a). The adductor sesamoid of the thumb. *American Journal of Physical Anthropology*, **20**, 297–302.
137. Garn, S. M. & Rohmann, C. G. (1962b). X-linked inheritance of developmental timing in man. *Nature*, **196**, 695–6.
138. Garn, S. M. & Rohmann, C. G. (1962c). Parent–child similarities in hand–wrist ossification. *American Journal of Disease of Children*, **103**, 603–7.
139. Garn, S. M. & Rohmann, C. G. (1964a). Compact bone deficiency in protein–calorie malnutrition. *Science*, **145**, 1444–5.

140. Garn, S. M. & Rohmann, C. G. (1964b). On the prevalence of skewness in incremental data. *American Journal of Physical Anthropology*, **21**, 235–6.
141. Garn, S. M. & Rohmann, C. G. (1966a). 'Communalities' in the ossification timing of the growing foot. *American Journal of Physical Anthropology*, **24**, 45–50.
142. Garn, S. M. & Rohmann, C. G. (1966b). Interaction of nutrition and genetics in the timing of growth and development. *Pediatric Clinics of North America*, **13**, 353–79.
143. Garn, S. M. & Rohmann, C. G. (1967). 'Midparent' values for use with parent-specific, age–size tables when paternal stature is estimated or unknown. *Pediatric Clinics of North America*, **14**, 283.
144. Garn, S. M., Rohmann, C. G. & Apfelbaum, B. (1961b). Complete epiphyseal union of the hand. *American Journal of Physical Anthropology*, **19**, 365–72.
145. Garn, S. M., Rohmann, C. G. & Blumenthal, T. (1966b). Ossification sequence polymorphism and sexual dimorphism in skeletal development. *American Journal of Physical Anthropology*, **24**, 101–15.
146. Garn, S. M., Rohmann, C. G., Blumenthal, T. & Kaplan, C. S. (1966a). Developmental communalities of homologous and non-homologous body joints. *American Journal of Physical Anthropology*, **25**, 147–51.
147. Garn, S. M., Rohmann, C. G., Blumenthal, T. & Silverman, F. N. (1967f). Ossification communalities of the hand and other body parts: Their implications to skeletal assessment. *American Journal of Physical Anthropology*, **27**, 75–82.
148. Garn, S. M., Rohmann, C. G. & Davis, A. A. (1963c). Genetics of hand–wrist ossification. *American Journal of Physical Anthropology*, **21**, 33–40.
149. Garn, S. M., Rohmann, C. G. & Hertzog, K. P. (1969a). Apparent influence of the X chromosome on timing of 73 ossification centers. *American Journal of Physical Anthropology*, **30**, 123–8.
150. Garn, S. M., Rohmann, C. G. & Nolan, P. H., Jr. (1963e). Aging and bone loss in a normal human population. *Excerpta Medica Foundation*, **57**, 90.
151. Garn, S. M., Rohmann, C. G. & Nolan, P., Jr. (1964c). The development and nature of bone changes during aging. In *Relations of Development and Aging*, ed. J. E. Birren, pp. 44–61. Springfield, Illinois: Charles C Thomas.
152. Garn, S. M., Rohmann, C. G. & Robinow, M. (1961c). Increments in hand–wrist ossification. *American Journal of Physical Anthropology*, **19**, 45–53.
153. Garn, S. M., Rohmann, C. G. & Silverman, F. N. (1965e). Missing ossification centers of the foot – inheritance and developmental meaning. *Annales de Radiologie*, **8**, 629–44.
154. Garn, S. M., Rohmann, C. G. & Silverman, F. N. (1967e). Radiographic standards for postnatal ossification and tooth calcification. *Medical Radiography and Photography*, **43**, 45–66.
155. Garn, S. M., Rohmann, C. G., Wagner, B. & Ascoli, W. (1967g). Continuing bone growth throughout life: A general phenomenon. *American Journal of Physical Anthropology*, **26**, 313–18.
156. Garn, S. M., Rohmann, C. G., Wagner, B. & Davila, G. H. (1970). Dynamics of change at the endosteal surface of tubular bones. In *Progress in Methods of*

*Bone Mineral Measurement*, ed. G. D. Whedon & J. R. Cameron, pp. 430–79. Washington, DC: US Government Printing Office.

157. Garn, S. M., Rohmann, C. G., Wagner, B., Davila, G. H. & Ascoli, W. (1969b). Population similarities in the onset and rate of adult endosteal bone loss. *Clinical Orthopaedics*, **65**, 51–60.
158. Garn, S. M., Rohmann, C. G. & Wallace, D. K. (1961d). Association between alternate sequences of hand–wrist ossification. *American Journal of Physical Anthropology*, **19**, 361–4.
159. Garn, S. M. & Saalberg, J. H. (1953). Sex and age differences in the composition of the adult leg. *Human Biology*, **25**, 144–53.
160. Garn, S. M. & Schwager, P. M. (1967). Age dynamics of persistent transverse lines in the tibia. *American Journal of Physical Anthropology*, **27**, 357–8.
161. Garn, S. M., Selby, S. & Crawford, M. R. (1956b). Skin reflectance studies in children and adults. *American Journal of Anthropology*, **14**, 101–17.
162. Garn, S. M., Selby, S. & Crawford, M. R. (1956c). Skin reflectance during pregnancy. *American Journal of Obstetrics and Gynecology*, **72**, 974–6.
163. Garn, S. M., Selby, S. & Young, R. (1954). Scalp thickness and the fat loss theory of balding. *Archives of Dermatology and Syphilology*, **70**, 601–8.
164. Garn, S. M. & Shamir, Z. (1958). *Methods for Research in Human Growth*. Springfield, Illinois: Charles C Thomas.
165. Garn, S. M., Silverman, F. N. & Davis, A. A. (1963a). Skin and gonadal dosages during investigative radiography. *American Journal of Physical Anthropology*, **21**, 561–8.
166. Garn, S. M., Silverman, F. N. & Davis, A. A. (1964d). Gonadal dosages in investigative radiography. *Science*, **143**, 1039.
167. Garn, S. M., Silverman, F. N. & Rohmann, C. G. (1964e). A rational approach to the assessment of skeletal maturation. *Annales de Radiologie*, **7**, 297–307.
168. Garn, S. M., Silverman, F. & Sontag, L. W. (1957b). X-ray protection in studies of growth and development. *American Journal of Physical Anthropology*, **15**, 452.
169. Garn, S. M., Swindler, D. R. & Kerewsky, R. S. (1966d). A canine 'field' in sexual dimorphism in tooth size. *Nature*, **212**, 1501–2.
170. Garn, S. M. & Wagner, B. (1969). The adolescent growth of the skeletal mass and its implications to mineral requirements. In *Adolescent Nutrition and Growth*, ed. F. P. Heald, pp. 139–61. New York: Appleton-Century-Crofts.
171. Garn, S. M., Wagner, B., Rohmann, C. G. & Ascoli, W. (1968h). Further evidence for continuing bone expansion. *American Journal of Physical Anthropology*, **28**, 219–21.
172. Garn, S. M. & Young, R. W. (1956). Concurrent fat loss and fat gain. *American Journal of Physical Anthropology*, **14**, 497–504.

Gindhart, P. S. (1969). The frequency of appearance of transverse lines in the tibia in relation to childhood illness. *American Journal of Physical Anthropology*, **31**, 17–22.
Gindhart, P. S. (1971). Growth of the tibia and radius studied longitudinally in normal children in relation to childhood disorders. PhD Thesis, University of Texas at Austin.

Gindhart, P. S. (1972). The effect of seasonal variation on long bone growth. *Human Biology*, **44**, 335–50.

Gindhart, P. S. (1973). Growth standards for the tibia and radius in children aged one month through eighteen years. *American Journal of Physical Anthropology*, **39**, 41–8.

*Goldstein, H. (1979). *The Design and Analysis of Longitudinal Studies*. London: Academic Press.

Goodson, C. S. & Jamison, P. (1987). Relative rate of maturation: A reaffirmation of its significant psychological effect. *Journal of Biosocial Science*, **19**, 73–88.

*Greulich, W. & Pyle, S. I. (1959). *Radiographic Atlas of Skeletal Development of the Hand and Wrist*, 2nd edn. Stanford: Stanford University Press.

Guo, S. (1986). The use of impedance in the estimation of body composition. *American Journal of Physical Anthropology*, **69**, 209.

Guo, S. (1988). An application of logistic regression to study tracking of overweight. *American Journal of Physical Anthropology*, **75**, 218.

Guo, S. (1990). A computer program for smoothing using Kernel estimation. *American Statistical Association*, Proceedings, 306–308.

Guo, S., Chumlea, W. C., Siervogel, R. M. & Roche, A. F. (in press). Longitudinal analysis of plasma lipid levels in children and young adults. *Proceedings of the Fifth Conference for Federally Supported Human Nutrition Research Units and Centers*, Washington, DC.

Guo, S., Roche, A. F. & Chumlea, W. C. (1989b). Predicting fat-free mass (FFM) and percent body fat (%BF) from anthropometry, resistance and reactance. *American Journal of Human Biology*, **1**, 135.

Guo, S., Roche, A. F., Chumlea, W. C., Miles, D. S. & Pohlman, R. A. (1987a). Body composition predictions from bioelectric impedance. *Human Biology*, **59**, 221–33.

Guo, S., Roche, A. F. & Houtkooper, L. (1989a). Fat-free mass in children and young adults from bioelectric impedance and anthropometric variables. *American Journal of Clinical Nutrition*, **50**, 435–43.

Guo, S., Roche, A. F. & Moore, W. M. (1988). Reference data for head circumference status and 1-month increments from 1 to 12 months of age. *Journal of Pediatrics*, **113**, 490–4.

Guo, S., Siervogel, R. M. & Roche, A. F. (1989c). Tracking in body mass index from 2 to 18 years: The FELS Longitudinal Study. *Clinical Research*, **37**, 963A.

Guo, S., Siervogel, R. M. & Roche, A. F. (1990). Confidence limits for least-squares Kernel estimates. *Annual Meeting of the American Statistical Association*, pp. 119.

Guo, S., Siervogel, R. M., Roche, A. F. & Chumlea, W. C. (in press). Mathematical modelling of human growth: A comparative study. *American Journal of Human Biology*.

Guo, S., Simon, R., Talmadge, J. E. & Klabansky, R. L. (1987b). An interactive computer program for the analysis of growth curves. *Computers and Biomedical Research*, **20**, 37–48.

Hamill, P. V. V., Drizd, T. A., Johnson, C. L., Reed, R. B. & Roche, A. F. (1977). NCHS growth curves for children. Birth–18 years, United States. *Vital and*

*Health Statistics*, National Center for Health Statistics, Series 11, No. 165. Washington, DC: US Government Printing Office.

Hamill, P. V. V., Drizd, T. A., Johnson, C. L., Reed, R. B., Roche, A. F. & Moore, W. M. (1979). Physical growth: National Center for Health Statistics Percentiles. *American Journal of Clinical Nutrition*, **32**, 607–29.

Harrison, G. G., Buskirk, E. R., Carter, J. E. L., Johnston, F. E., Lohman, T. G., Pollock, M. L., Roche, A. F. & Wilmore, J. (1988). Skinfold thicknesses and measurement technique. In *Anthropometric Standardization Reference Manual*, ed. T. Lohman, A. F. Roche & R. Martorell, pp. 55–70. Champaign, Illinois: Human Kinetics Books.

Heiber, R. G. (1975). *The relationship of the ulnar sesamoid bone in males to circumpuberal growth rates of the mandible and body height and weight*. MSc Thesis, Ohio State University.

Hertzog, K. P. (1967). Shortened fifth middle phalanges. *American Journal of Physical Anthropology*, **27**, 113–18.

Hertzog, K. P. (1990). Evidence for continuing bone elongation in adults based on longitudinal hand radiographs. *Acta medica Auxologica*, **22**, 57–60.

Hertzog, K. P., Garn, S. M. & Church, S. F. (1968). Cone-shaped epiphyses in the hand. *Investigative Radiology*, **3**, 433–41.

Hertzog, K. P., Garn, S. M. & Hempy, H. O., III. (1969). Partitioning the effects of secular trend and aging on adult stature. *American Journal of Physical Anthropology*, **31**, 111–15.

Himes, J. H. (1977). Gruelich–Pyle and Tanner–Whitehouse skeletal age: Associations with pubertal events. *American Journal of Physical Anthropology*, **47**, 136–7.

Himes, J. H. (1978a). Bone growth and development in protein–calorie malnutrition. *World Review of Nutrition and Dietetics*, **28**, 143–87.

Himes, J. H. (1978b). Infant feeding patterns and subsequent obesity. *American Journal of Physical Anthropology*, **48**, 405–6.

Himes, J. H. (1979). Secular trends in body proportions and composition. In *Secular Trends in Child Growth, Maturation and Development*, ed. A. F. Roche, pp. 28–59, Monographs of the Society for Research in Child Development, **44**.

Himes, J. H. (1980a). Subcutaneous fat thickness as an indicator of nutritional status. In *Social and Biological Predictors of Nutritional Status, Physical Growth and Neurological Development*, ed. L. S. Greene & F. E. Johnston, pp. 3–26. New York: Academic Press.

Himes, J. H. (1980b). Skinfold compression and the estimation of total body fat. *American Journal of Physical Anthropology*, **52**, 237–8.

Himes, J. H. (1984a). Appropriateness of parent-specific stature adjustment for US black children. *Journal of the National Medical Association*, **76**, 55–7.

Himes, J. H. (1984b). Appropriateness of the linear midparent model for parent–child stature studies. *American Journal of Physical Anthropology*, **63**, 170.

Himes, J. H. & Roche, A. F. (1982). Reported versus measured adult statures. *American Journal of Physical Anthropology*, **58**, 335–42.

Himes, J. H. & Roche, A. F. (1986). Subcutaneous fatness and stature: Relationships from infancy to adulthood. *Human Biology*, **58**, 737–50.

Himes, J. H., Roche, A. F. & Siervogel, R. M. (1979). Compressibility of skinfolds and the measurement of subcutaneous fatness. *American Journal of Clinical Nutrition*, **32**, 1734–40.

Himes, J. H., Roche, A. F. & Thissen, D. (1981). Parent-specific adjustments for assessment of recumbent length and stature. *Monographs in Pediatrics*, **13**, 1–88.

Himes, J. H., Roche, A. F., Thissen, D. & Moore, W. M. (1985). Parent-specific adjustments for evaluation of recumbent length and stature. *Pediatrics*, **75**, 304–13.

Himes, J. H., Roche, A. F. & Webb, P. (1980). Fat areas as estimates of total body fat. *American Journal of Clinical Nutrition*, **33**, 2093–100.

Hinck, V. C. & Hopkins, C. E. (1965). Concerning growth of the sphenoid sinus. *Archives of Otolaryngology*, **82**, 62–6.

*Hollingshead, A. (1957). *The Two Factor Index of Social Position*. Atlanta, Georgia: Emory University.

Holm, V. A., Kronmall, R. A., Williamson, M. & Roche, A. F. (1979). Physical growth in phenylketonuria: II. Growth of children in the PKU collaborative study from birth to four years of age. *Pediatrics*, **63**, 700–7.

Huff, C. E. & Pyle, S. I. (1937). Differences in the utilization of calcium and phosphorus in negative and positive balances during pregnancy. *Human Biology*, **9**, 29–42.

Israel, H. (1966). Radiogrammetric replicability and direct validity of small mandibular measurements in the 45-degree projection. *Journal of Dental Research*, **45**, 1570.

Israel, H. (1967a). Loss of bone and remodeling–redistribution in the craniofacial skeleton with age. *Federation Proceedings*, **26**, 1723–8.

Israel, H. (1967b). Microdensitometric analysis for the study of cranio-facial growth in the living subject. *American Journal of Physical Anthropology*, **27**, 236.

Israel, H. (1968a). Continuing growth in the human cranial skeletal. *Archives of Oral Biology*, **13**, 133–8.

Israel, H. (1968b). *In vivo* assessment of skeletal morphology by radiographic densitometry. *Journal of Periodontology*, **38**, 667–76.

Israel, H. (1969). Pubertal influence upon the growth and sexual differentiation of the human mandible. *Archives of Oral Biology*, **14**, 583–90.

Israel, H. (1970). Continuing growth in sella turcia with age. *American Journal of Roentgenology*, **108**, 516–27.

Israel, H. (1971). The impact of aging upon the adult craniofacial skeleton. PhD Dissertation, University of Alabama at Birmingham.

Israel, H. (1973a). Age factor and the pattern of change in craniofacial structures. *American Journal of Physical Anthropology*, **39**, 111–28.

Israel, H. (1973b). Progressive enlargement of the vertebral body as part of the process of human skeletal aging. *Age and Aging*, **2**, 71–9.

Israel, H. (1973c). Recent knowledge concerning craniofacial aging. *Angle Orthodontist*, **43**, 176–84.

Israel, H. (1973d). The failure of aging or loss of teeth to drastically alter mandibular angle morphology. *Journal of Dental Research*, **52**, 83–90.

Israel, H. (1977). The dichotomous pattern of craniofacial expansion during aging. *American Journal of Physical Anthropology*, **47**, 47–52.

Israel, H. (1978). The fundamentals of cranial and facial growth. In *Human Growth*, vol. 2, ed. F. Falkner & J. M. Tanner, pp. 357–80. New York: Plenum Press.

Israel, H., Garn, S. M. & Colbert, C. (1967). The recording microdensitometer as a research tool in the study of oral–facial development. *Journal of Dental Research*, **46**, 164.

Israel, H. & Lewis, A. B. (1971). Radiographically determined linear permanent tooth growth from age 6 years. *Journal of Dental Research*, **50**, 334–42.

Johnson, G. F., Dorst, J. P., Kuhn, J. P., Roche, A. F. & Davila, G. H. (1973). Reliability of skeletal age assessments. *American Journal of Roentgenology*, **158**, 320–7.

Johnston, F. E., Roche, A. F., Schell, L. M. & Wettenhall, H. N. B. (1975). Critical weight at menarche. Critique of an hypothesis. *American Journal of Diseases of Children*, **129**, 19–23.

Johnston, F. E., Roche, A. F. & Susanne, C. (1980). *Human Physical Growth and Maturation: Methodologies and Factors*. New York: Plenum Press.

*Jones, M. C., Bayley, N., Macfarlane, J. W. & Honzik, M. P. (1971). *The Course of Human Development*. Waltham, Massachusetts: Xerox College Publishing Co.

*Jordan, J., Ruben, M., Hernandez, J., Bebelagua, A., Tanner, J. M. & Goldstein, H. (1975). The 1972 Cuban National Growth Study as an example of population health monitoring: design and methods. *Annals of Human Biology*, **2**, 153–71.

Kagan, J. & Garn, S. M. (1963). A constitutional correlate of early intellective functioning. *Journal of Genetic Psychology*, **102**, 83–9.

Kagan, J. & Moss, H. A. (1963). *Birth to Maturity: A Study in Psychological Development*. New York: Wiley.

Kingsley, A. & Reynolds, E. L. (1949). The relation of illness patterns in children to ordinal position in the family. *Journal of Pediatrics*, **35**, 17–23.

Koski, K. (1961). Growth changes in the relationships between some basicranial planes and palatal plane. *Suomen Hammaslaakariseuran Toimituksia*, **57**, 15–26.

Koski, K. & Garn, S. M. (1957). Tooth eruption sequence in fossil and modern man. *American Journal of Physical Anthropology*, **15**, 469–88.

Koski, K., Garn, S. M. & Lewis, A. B. (1957). Tooth eruption sequence in fossil and modern man. *American Journal of Physical Anthropology*, **15**, 451–2.

Kouchi, M., Mukherjee, D. & Roche, A. F. (1985a). Curve fitting for growth in weight during infancy with relationships to adult status, and familial associations of the estimated parameters. *Human Biology*, **57**, 245–65.

Kouchi, M., Roche, A. F. & Mukherjee, D. (1985b). Growth in recumbent length during infancy with relationships to adult status and familial associations of the estimated parameters. *Human Biology*, **57**, 449–72.

*Kushner, R. F. & Schoeller, D. A. (1986). Estimation of total body water by bioelectrical impedance analysis. *American Journal of Clinical Nutrition*, **44**, 417–24.

Lavelle, M. (1991). Predictors of small size of the human birth canal. *American Journal of Physical Anthropology*, **12**, 112.

Lee, M. M. C. (1967). Natural markers in bone growth. *American Journal of Physical Anthropology*, **27**, 237.

Lee, M. M. C. & Garn, S. M. (1967). Pseudoepiphyses or notches in the non-epiphyseal end of metacarpal bones in healthy children. *Anatomical Record*, **159**, 263–72.

Lee, M. M. C., Garn, S. M. & Rohmann, C. G. (1968). Relation of metacarpal notching to stature and maturational status of normal children. *Investigative Radiology*, **3**, 96–102.

Lee, P. S. T. & Guo, S. (1986). A curve fitting analysis of the normal approximation to binomial distribution. In *Program and Abstracts*, p. 170, Chicago: Joint Statistical Meetings. American Statistical Association; Biometric Society, Institute of Mathematical Statistics.

Lestrel, P. E. & Brown, H. D. (1976). Fourier analysis of adolescent growth of the cranial vault: A longitudinal study. *Human Biology*, **48**, 517–28.

Lestrel, P. E., Engstrom, C. & Bodt, A. (1991). Quantitative analysis of nasal bone growth: The first year of life. *American Journal of Physical Anthropology*, **112**, 114–15.

*Lestrel, P. E. & Roche, A. F. (1977). A comparative study of cranial thickness in Down's syndrome: Fourier analysis. *Acta Medica Auxologica*, **9**, 27.

Lestrel, P. E. & Roche, A. F. (1984). Variability in cranial base shape with age: Fourier analysis. *American Journal of Physical Anthropology*, **63**, 184.

Lestrel, P. E. & Roche, A. F. (1986). Cranial base shape variation with age: A longitudinal study of shape using Fourier analysis. *Human Biology*, **58**, 527–40.

Lewis, A. B. & Garn, S. M. (1960). The relationship between tooth formation and other maturation factors. *Angle Orthodontist*, **30**, 70–7.

Lewis, A. B. & Roche, A. F. (1974). Cranial base elongation in boys during pubescence. *Angle Orthodontist*, **44**, 83–93.

Lewis, A. B. & Roche, A. F. (1977). The saddle angle: Constancy or change? *Angle Orthodontist*, **47**, 46–54.

Lewis, A. B. & Roche, A. F. (1988). Late growth changes in the craniofacial skeleton. *Angle Orthodontist*, **58**, 127–35.

Lewis, A. B., Roche, A. F. & Wagner, B. (1982). The growth of the mandible during pubescence. *Angle Orthodontist*, **52**, 325–42.

Lewis, A. B., Roche, A. F. & Wagner, B. (1985). Pubertal spurts in cranial base and mandible: Comparisons within individuals. *Angle Orthodontist*, **55**, 17–30.

*Lohman, T. G. (1986). Applicability of body composition techniques and constants for children and youths. *Exercise and Sport Science Reviews*, **14**, 325–57.

Lohman, T. G., Roche, A. F. & Martorell, R. (1988). *Anthropometric Standardization Reference Manual*. Champaign, Illinois: Human Kinetics Books.

Malina, R. M. & Roche, A. F. (1983). *Manual of Physical Status and Performance in Childhood*. Vol. II, *Physical Performance*. New York: Plenum Press.

McCall, R. B., Meyers, E. D., Jr., Hartman, J. & Roche, A. F. (1983). Development changes in head-circumference and mental-performance growth rates: A test of Epstein's prenoblysis hypothesis. *Developmental Psychobiology*, **16**, 457–68.

*McCammon, R. W. (1970). *Human Growth and Development*. Springfield, Illinois: Charles C Thomas.

Meier, R. J., Goodson, C. S. & Roche, E. (1987). Dermatoglyphic development and timing of maturation. *Human Biology*, **59**, 357–73.

Moerman, M. L. (1981). A longitudinal study of growth in relation to body size and sexual dimorphism in the human pelvis. PhD Thesis, University of Michigan: Ann Arbor, Michigan.

Moerman, M. L. (1982). Growth of the birth canal in adolescent girls. *American Journal of Obstetrics and Gynecology*, **143**, 528–32.

Moore, W. M. & Roche, A. F. (1982). *Pediatric Anthropometry*. Columbus, Ohio: Ross Laboratories.

Moore, W. M. & Roche, A. F. (1983). *Pediatric Anthropometry*, 2nd edn. Columbus, Ohio: Ross Laboratories.

Moore, W. M. & Roche, A. F. (1987). *Pediatric Anthropometry*, 3rd edn. Columbus, Ohio: Ross Laboratories.

Mukherjee, D. (1982). Fitting distributions to non-normal data. *American Journal of Physical Anthropology*, **57**, 212.

Mukherjee, D. & Hurst, D. C. (1984). Maximum entropy revisited. *Statistica Neerlandica*, **38**, 1–13.

Mukherjee, D., Neiswanger, K., Siervogel, R. M., Roche, A. F. & Roche, E. (1984). Genetic analyses of discrete or continuous traits upon fitting a finite mixture of flexible maximum entrophy distributions. *American Journal of Human Genetics*, **36**, 176S.

Mukherjee, D. & Roche, A. F. (1984). The estimation of percent body fat, body density and total body fat by maximum $R^2$ regression equations. *Human Biology*, **56**, 79–109.

Mukherjee, D. & Siervogel, R. M. (1983). An alternative to the mixture of normal distributions. *American Journal of Human Genetics*, **35**, 202A.

Murray, J. R., Bock, R. D. & Roche, A. F. (1971). The measurement of skeletal maturity. *American Journal of Physical Anthropology*, **35**, 327–30.

Newbery, H. (1941). Studies in fetal behavior. IV. The measurement of three types of fetal activity. *Journal of Comparative Psychology*, **32**, 521–30.

Norman, H. N. (1942). Fetal hiccups. *Journal of Comparative Psychology*, **34**, 65–73.

Ohtsuki, F., Mukherjee, D., Lewis, A. B. & Roche, A. F. (1982a). A factor analysis of cranial base and vault dimensions in children. *American Journal of Physical Anthropology*, **58**, 271–9.

Ohtsuki, F., Mukherjee, D., Lewis, A. B. & Roche, A. F. (1982b). Growth of cranial base and vault dimensions in children. *Journal Anthropological Society of Nippon*, **90**, 239–58.

Olson, J. M., Boehnke, M., Neiswanger, K., Roche, A. F. & Siervogel, R. M. (1989). Alternative genetic models for the inheritance of the phenylthiocarbamide taste deficiency. *Genetic Epidemiology*, **6**, 423–34.

*Palmer, C. E. & Reed, L. J. (1935). Anthropometric studies of individual growth. I. Age, height and growth in height; elementary school children. *Human Biology*, **7**, 319–24.

Pao, E. M., Himes, J. H. & Roche, A. F. (1980). Milk intake and feeding patterns of breast-fed infants. *Journal of American Dietetic Association*, 77, 540–5.

Patton, J. L. (1979). A study of distributional normality of skinfold measurements. Master's Thesis. Seattle, Washington: University of Washington.

Potter, D. E., Broyer, M., Chantler, C., Gruskin, A., Holliday, M. A., Roche, A. F., Scharer, K. & Thissen, D. (1978). The measurement of growth in children with renal insufficiency. *Kidney International*, 14, 378–82.

Poznanski, A. K., Garn, S. M., Nagy, J. M. & Gall, J. C., Jr. (1972). Metacarpo-phalangeal pattern profiles in the evaluation of skeletal malformation. *Radiology*, 104, 1–11.

Poznanski, A. K., Roche, A. F., Mukherjee, D., Pachman, L. M. & Brewer, E. J., Jr. (1985). Norms of the apparent width of the knee joint: Useful measures in the evaluation of children with juvenile arthritis. *American Journal of Roentgenology, Radium Therapy and Nuclear Medicine*, 145, 870.

Pyle, S. I. (1938). Interrelations of hemoglobin, basal metabolism, and creatine, creatinine and magnesium excretion during human pregnancy. *Human Biology*, 10, 528–36.

Pyle, S. I. (1939). Observations on the size and position of the nutrient foramen in the radius. *Human Biology*, 2, 369–78.

Pyle, S. I. & Huff, C. E. (1936). The use of 3-day periods in the human metabolism studies. Calcium and phosphorus. *Journal of Nutrition*, 11, 495–509.

Pyle, S. I. & Menino, C. (1939). Observations on estimating skeletal age from the Todd and Flory bone atlases. *Child Development*, 10, 27–34.

Pyle, S. I., Potgieter, M. & Comstock, G. (1938). On certain relationships of calcium in the blood serum to calcium balance and basal metabolism during pregnancy. *American Journal of Obstetrics and Gynecology*, 35, 283–9.

Pyle, S. I. & Sontag, L. W. (1943). Variability in onset of ossification in epiphyses and short bones of the extremities. *American Journal of Roentgenology and Radium Therapy*, 49, 795–8.

Reynolds, E. L. (1943). Degree of kinship and pattern of ossification. *American Journal of Physical Anthropology*, 1, 405–16.

Reynolds, E. L. (1944). Differential tissue growth in the leg during childhood. *Child Development*, 15, 181–205.

Reynolds, E. L. (1945). Bony pelvic girdle in early infancy. *American Journal of Physical Anthropology*, 3, 321–54.

Reynolds, E. L. (1946a). Sexual maturation and the growth of fat, muscle and bone in girls. *Child Development*, 17, 121–44.

Reynolds, E. L. (1946b). The bony pelvis in prepubertal childhood. *American Journal of Physical Anthropology*, 5, 165–200.

Reynolds, E. L. (1948). Distribution of tissue components in the female leg from birth to maturity. *Anatomical Record*, 100, 621–30.

Reynolds, E. L. (1949a). Anthropology and human growth. *Ohio Journal of Science*, 49, 89–91.

Reynolds, E. L. (1949b). The fat/bone index as a sex-differentiating character in man. *Human Biology*, 21, 199–204.

Reynolds, E. L. (1951). The distribution of subcutaneous fat in childhood and adolescence. *Monographs of the Society for Research in Child Development*, No. 50.

Reynolds, E. L. & Asakawa, T. (1948). Measurement of obesity in childhood. *American Journal of Physical Anthropology*, **6**, 475–87.

Reynolds, E. L. & Asakawa, T. (1950). A comparison of certain aspects of body structure and body shape in 200 adults. *American Journal of Physical Anthropology*, **8**, 343–65.

Reynolds, E. L. & Asakawa, T. (1951). Skeletal development in infancy. *American Journal of Roentgenology and Radium Therapy*, **65**, 403–10.

Reynolds, E. L. & Clark, L. C., Jr. (1947). Creatinine excretion, growth progress and body structure in normal children. *Child Development*, **18**, 155–68.

Reynolds, E. L. & Grote, P. (1948). Sex differences in the distribution of tissue components in the human leg from birth to maturity. *Anatomical Record*, **102**, 45–53.

Reynolds, E. L. & Schoen, G. (1947). Growth patterns of identical triplets from 8 through 18 years. *Child Development*, **18**, 130–51.

Reynolds, E. L. & Sontag, L. W. (1944). Seasonal variations in weight, height and appearance of ossification centers. *Journal of Pediatrics*, **24**, 524–35.

Reynolds, E. L. & Sontag, L. W. (1945). Variations in growth patterns in health and disease. *Journal of Pediatrics*, **26**, 336–52.

Reynolds, E. L. & Wines, J. (1948). Individual differences in physical changes associated with adolescence in girls. *American Journal of Diseases of Children*, **75**, 329–50.

Reynolds, E. L. & Wines, J. V. (1951). Physical changes associated with adolescence in boys. *American Journal of Diseases of Children*, **82**, 529–47.

Richards, T. W. & Nelson, V. L. (1938). Studies in mental development: II. Analysis of abilities tested at the age of six months by the Gesell schedule. *Journal of Genetic Psychology*, **52**, 327–31.

Richards, T. W. & Newbery, H. (1938). Studies in fetal behavior: III. Can performance on test items at six months postnatally be predicted on the basis of fetal activity? *Child Development*, **9**, 79–86.

Richards, T. W., Newbery, H. & Fallgatter, R. (1938). Studies in fetal behavior: II. Activity of the human fetus in utero and its relation to other prenatal conditions, particularly the mother's basal metabolic rate. *Child Development*, **9**, 69–78.

Robinow, M. (1942a). The variability of weight and height increments from birth to six years. *Child Development*, **13**, 159–64.

Robinow, M. (1942b). Appearance of ossification centers. Groupings obtained from factor analysis. *American Journal of Diseases of Children*, **64**, 229–36.

Robinow, M. (1943). The statistical diagnosis of zygosity in multiple human births. *Human Biology*, **15**, 221–35..

Robinow, M. (1968). Field measurement of growth and development. In *Malnutrition, Learning and Behavior*, ed. N. S. Scrimshaw & J. E. Gordon, pp. 409–25. Cambridge, Massachusetts: Massachusetts Institute of Technology Press.

Robinow, M., Johnson, M. & Anderson, M. (1943a). Feet of normal children. *Journal of Pediatrics*, **23**, 141–9.

Robinow, M., Leonard, V. L. & Anderson, M. (1943b). A new approach to the analysis of children's posture. *Journal of Pediatrics*, **22**, 655–63.

Robinow, M., Richards, T. W. & Anderson, M. (1942). The eruption of deciduous teeth. *Growth*, **6**, 127–33.
173. *Roche, A. F. (1965). The sites of elongation of human metacarpals and metatarsals. *Acta Anatomica (Basel)*, **61**, 193–202.
174. Roche, A. F. (1971a). New statistical technique that assist the interpretation of serial growth data. In *Growth and Development*, pp. 9–11. Proceedings XIII International Congress of Pediatrics, Vienna.
175. Roche, A. F. (1971b). Summary of discussion. *American Journal of Physical Anthropology*, **35**, 467–70.
176. Roche, A. F. (1971c). Symposium on Assessment of Skeletal Maturity. *American Journal of Physical Anthropology*, **35**, 315–469.
177. Roche, A. F. (1973). The timing and sequence of some adolescent events. In *Report on Workshop on Trauma in Adolescents in Sports and Recreation*, pp. 1–40. Committee on Prosthetics Research and Development. Washington, DC: National Academy of Science.
178. Roche, A. F. (1974a). Differential timing of maximum length increments between bones within individuals. *Human Biology*, **46**, 145–57.
179. Roche, A. F. (1974b). The design of studies for measuring growth. *Proceedings of XIV International Congress of Pediatrics*, **58**, 320–31.
180. Roche, A. F. (1974c). Introduction to symposium: Some aspects of adolescent physiology. *Human Biology*, **46**, 115–16.
181. Roche, A. F. (1974d). Anthropometric indices as nutritional indicators. *Proceedings of XIV International Congress of Pediatrics*, **5**, 1–12.
182. Roche, A. F. (1975). Some aspects of adolescent growth and maturation. In *Nutrient Requirements in Adolescence*, ed. H. N. Munroe & J. I. McKigney, pp. 33–56. Cambridge, Massachusetts: Massachusetts Institute of Technology Press.
183. Roche, A. F. (1976). Growth after puberty. In *Youth in a Changing World: Cross-cultural Perspectives on Adolescence*, ed. E. E. Fuchs, pp. 17–53. The Hague: Mouton Publishers.
184. Roche, A. F. (1978a). Growth assessment in abnormal children. *Kidney International*, **14**, 369–77.
185. Roche, A. F. (1978b). Parametros de crecimiento. *Archivos Argentinos de Pediatria*, **76**, 8–13+52.
186. Roche, A. F. (1978c). Bone growth and maturation. In *Human Growth*, ed. F. Falkner & J. M. Tanner, pp. 318–55. New York: Plenum Publishing Corporation.
187. Roche, A. F., ed. (1979a). *Secular Trends in Human Growth, Maturation and Development*. Monographs of the Society for Research in Child Development, **44**.
188. Roche, A. F. (1979b). Postnatal growth of adipose tissue in man. *Studies in Physical Anthropology*, **5**, 53–73.
189. Roche, A. F. (1979c). Growth assessment of handicapped children. *Dietetic Currents*, **6**, 25–30.
190. Roche, A. F. (1980a). The analysis of serial data. *Studies in Physical Anthropology*, **6**, 71–88.
191. Roche, A. F. (1980b). Prediction. In *Human Physical Growth and Maturation: Methodologies and Factors*, ed. F. E. Johnston, A. F. Roche & C. Susanne,

pp. 177–91. NATO Advanced Study Institute, Sogesta, Italy. New York: Plenum Publishing Corporation.

192. Roche, A. F. (1980c). Possible catch-up growth of the brain. *Acta Medica Auxologica*, **12**, 165–79.

193. Roche, A. F. (1981a). Recent advances in child growth and development. Eleventh International Congress of Anatomy. In *Glial and Neuronal Cell Biology*, ed. S. Fedoroff, pp. 321–9. New York: Alan R. Liss.

194. Roche, A. F. (1981b). The adipocyte number hypothesis. *Child Development*, **52**, 31–43.

195. Roche, A. F. (1982). Anthropometric variables: Effectiveness and limitations. In *Assessing the Nutritional Status of the Elderly – State of the Art*, ed. C. W. Calloway & G. Harrison, pp. 22–8. Report of the Third Ross Roundtable on Medical Issues. Columbus, Ohio: Ross Laboratories.

196. Roche, A. F. (1984a). Research progress in the field of body composition. *Medicine and Science in Sports and Exercise*, **16**, 579–83.

197. Roche, A. F. (1984b). Adult stature prediction: A critical review. *Acta Medica Auxologica*, **16**, 5–28.

198. Roche, A. F. (1984c). Anthropometric methods: New and old, what they tell us. *International Journal of Obesity*, **8**, 509–23.

199. Roche, A. F. (1985a). Concluding remarks. In *Body Composition Assessments in Youth and Adults*, ed. A. F. Roche, pp. 107–8. Proceedings of Sixth Ross Conference on Medical Research, Williamsburg, Virginia, December 1984. Columbus, Ohio: Ross Laboratories.

200. Roche, A. F. (1985b). Continuities and discontinuities in postnatal growth. *Book of Abstracts*, p. 601. XII International Anatomical Congress 1985, London, Barbican Centre.

201. Roche, A. F. (1986a). Physical growth. In *An Evaluation and Assessment of the State of the Science. Report of the Study Group on Developmental Endocrinology and Physical Growth*, ed. S. L. Kaplan & G. D. Grave, pp. 43–73. Washington, DC: US Department of Health and Human Services.

202. Roche, A. F. (1986b). Bone growth and maturation. In *Human Growth. A Comprehensive Treatise*, 2nd edn., vol. 2, *Postnatal Growth, Neurobiology*, ed. F. Falkner & J. M. Tanner, pp. 25–60. New York: Plenum Press.

203. Roche, A. F. (1986c). The need for improvements in the measurements of body composition: Some critical issues. *American Journal of Physical Anthropology*, **69**, 256.

204. Roche, A. F. (1987a). Some aspects of the criterion methods for the measurement of body composition. *Human Biology*, **59**, 209–20.

205. Roche, A. F. (1987b). Skeletal status in normal children. Report of the Seventh Ross Conference on Medical Research. *Osteoporosis: Current Concepts*, ed. A. F. Roche, pp. 8–11. Columbus, Ohio: Ross Laboratories.

206. Roche, A. F. (1987c). Risk of overweight at 18 years dependent on weight and weight/stature$^2$ during childhood. *Clinical Research*, **35**, 907a.

207. Roche, A. F. (1989a). The final phase of growth in stature. *Growth, Genetics and Hormones*, **5**, 4–6.

208. Roche, A. F. (1989b). Infants, children, and adolescents. In *Nutritional Status Assessments of the Individual*, ed. G. E. Livingston, pp. 179–87. Trumbull, Connecticut: Food and Nutrition Press.

209. Roche, A. F. (1989c). Relative utility of carpal skeletal ages. *American Journal of Human Biology*, **1**, 479–82.
210. Roche, A. F., Abdel-Malek, A. D. & Mukherjee, D. (1985). New approaches to clinical assessment of adipose tissue. In *Body Composition Assessments in Youth and Adults*, ed. A. F. Roche, pp. 14–19. Proceedings of Sixth Ross Conference on Medical Research, Williamsburg, Virginia, December 1984. Columbus, Ohio, Ross Laboratories.
211. Roche, A. F. & Baumgartner, R. N. (1988). Tracking in fat distribution during growth. In *Fat Distribution During Growth and Later Health Outcomes*, ed. C. Bouchard & F. E. Johnston, pp. 147–62. New York: Alan R. Liss.
212. Roche, A. F., Baumgartner, R. N. & Guo, S. (1987a). Population methods: Anthropometry or estimations. In *Human Body Composition and Fat Distribution*, ed. N. G. Norgan. *Euronut Report*, **8**, 31–48.
213. Roche, A. F., Baumgartner, R. N. & Guo, S. (1991a). Anthropometry: Classical and modern approaches. In *New Techniques in Nutritional Research*, ed. R. G. Whitehead & A. Prentice, pp. 241–59. San Diego, CA: Academic Press.
214. Roche, A. F., Baumgartner, R. N. & Siervogel, R. M. (1991b). B-mode ultrasound measurement of subcutaneous adipose tissue. *American Journal of Clinical Nutrition*, **53**, 27.
215. Roche, A. F. & Chumlea, W. C. (1980). Serial changes in predicted adult statures for individuals. *Human Biology*, **52**, 507–13.
216. Roche, A. F. & Chumlea, W. C. (in press). New approaches to clinical assessment of adipose tissue. In *Obesity*, ed. B. N. Brodoff & P. Bjorntorp. New York: J. B. Lippincott.
217. Roche, A. F., Chumlea, W. C. & Guo, S. (1986a). *Identification and Validation of New Anthropometric Techniques for Quantifying Body Composition*. Technical Report Natick/TR-86/058. Natick, Massachusetts: United States Army Natick Research, Development and Engineering Center.
218. Roche, A. F., Chumlea, W. C. & Guo, S. (1987c). Estimation of body composition from impedance. *Medicine and Science in Sports and Exercise*, **19**, S40.
219. Roche, A. F., Chumlea, W. C. & Siervogel, R. M. (1982b). *Longitudinal Study of Human Hearing: Its Relationship to noise and other factors. III. Results from the first five years*, AMRL-TR-82-68, pp. 216. Wright-Patterson Air Force Base, Ohio: Aerospace Medical Research Laboratory.
220. Roche, A. F., Chumlea, W. C., Siervogel, R. M. & Mukherjee, D. (1983d). *Lonitudinal Study of Human Hearing: Its relationship to noise and other factors. IV. Data from 1976 to 1982. Final Report*, AMRL-TR-83-057, pp. 62. Wright-Patterson Air Force Base, Ohio: Aerospace Medical Research Laboratory.
221. Roche, A. F., Chumlea, W. C. & Thissen, D. (1988a). *Assessing the Skeletal Maturity of the Hand–Wrist: FELS Method*, pp. viii + 339. Springfield, Illinois: Charles C Thomas.
222. Roche, A. F. & Davila, G. H. (1972). Late adolescent growth in stature. *Pediatrics*, **50**, 874–80.
223. Roche, A. F. & Davila, G. H. (1974a). Differences between recumbent length and stature within individuals. *Growth*, **38**, 313–20.

224. Roche, A. F. & Davila, G. H. (1974b). Growth after puberty. *Proceedings, XIV International Congress of Pediatrics, Buenos Aires,* **5**, 138–52.
225. Roche, A. F. & Davila, G. H. (1975). Prepubertal and postpubertal growth period. In *Fetal and Postnatal Growth – Hormones and Nutrition,* ed. D. B. Cheek, pp. 409–14. New York: John Wiley.
226. Roche, A. F. & Davila, G. H. (1976). The reliability of assessments of the maturity of individual hand–wrist bones. *Human Biology,* **48**, 585–97.
227. Roche, A. F., Davila, G. H. & Mellits, E. D. (1975c). Late adolescent changes in weight. In *Biosocial Interrelations in Population Adaptation,* ed. E. S. Watts, F. E. Johnston & C. W. Lasker, pp. 309–18. The Hague: Mouton Publishers.
228. Roche, A. F., Davila, G. H., Pasternack, B. A. & Walton, M. J. (1970a). Some factors influencing the replicability of assessments of skeletal maturity (Greulich–Pyle). *American Journal of Roentgenology,* **109**, 299–306.
229. Roche, A. F., Eichorn, D., McCammon, R. W., Reed, R. B., Valadian, I., Himes, J. H., Kent, R. L., Jr. & Siervogel, R. M. (1980). *The Natural History of Blood Pressure.* Report on Contract N01-HV-42985, 1400 pp.
230. Roche, A. F., Eyman, S. L. & Davila, G. H. (1971). Skeletal age prediction. *Journal of Pediatrics,* **78**, 997–1003.
231. Roche, A. F. & Falkner, F., eds. (1974). *Nutrition and Malnutrition: Identification and Measurement.* New York: Plenum Press.
232. Roche, A. F. & Falkner, F. (1975). Physical growth charts. In *Pediatric Screening Tests,* ed. W. K. Frankenburg & B. W. Camp, pp. 63–73. Springfield, Illinois: Charles C Thomas.
233. Roche, A. F. & French, N. Y. (1969). Rapid changes in weight and growth potential during childhood. *American Journal of Physical Anthropology,* **31**, 231–33.
234. Roche, A. F. & French, N. Y. (1970). Differences in skeletal maturity levels between the knee and hand. *American Journal of Roentgenology,* **109**, 307–12.
235. Roche, A. F., Garn, S. M., Reynolds, E. L., Robinow, M. & Sontag, L. W. (1981a). The first seriatim study of human growth and middle aging. *American Journal of Physical Anthropology,* **54**, 23–24.
236. Roche, A. F. & Guo, S. (1987). Tracking of weight (W) and weight/stature$^2$ (W/S$^2$) from one month to 30 years. *International Journal of Obesity,* **11**, 436A.
237. Roche, A. F. & Guo, S. (1988). Review of the paper by Cole, T. J.: Fitting smoothed centile curves to reference data. *Journal of the Royal Statistical Society,* **151**, 414.
238. Roche, A. F. & Guo, S. (1990). Prediction of fat-free mass (FFM) from bioelectric impedance and anthropometry. *Clinical Research,* **38**, 756A.
239. Roche, A. F., Guo, S. & Baumgartner, R. N. (1988b). The measurement of stature. *American Journal of Clinical Nutrition,* **47**, 922.
240. Roche, A. F., Guo, S. & Chumlea, W. C. (1989d). *Equations to Predict Fat-free Mass in Adults.* Report to RJL Systems, Inc. Detroit, Michigan.
241. Roche, A. F., Guo, S. & Houtkooper, L. (1989b). Biased estimation of fat-free mass. American Statistical Association. *Proceedings of the Biopharmaceutical Section,* pp. 188–91. 51st Annual Meeting, New Orleans, Louisiana.
242. Roche, A. F., Guo, S. & Moore, W. M. (1989a). Weight and recumbent

length from 1 to 12 months of age: Reference data for 1-month increments. *American Journal of Clinical Nutrition*, **49**, 599–607.

243. Roche, A. F., Guo, S. & Moore, W. M. (1989c). Infant growth and breast-feeding. *American Journal of Clinical Nutrition*, **50**, 1117–18.

244. Roche, A. F., Guo, S. & Moore, W. M. (1991). Body surface area of infants: reference data. *American Journal of Physical Anthropology*, **12**, 152.

245. Roche, A. F. & Hamill, P. V. V. (1977). United States Growth Charts. In *Proceedings of the First International Congress of Auxology, Rome*, pp. 143–8.

246. Roche, A. F. & Hamill, P. V. V. (1978). United States Growth Charts. In *Auxology: Human Growth in Health and Disorder*, ed. L. Gedda & P. Parisi, pp. 133–8. London, England: Academic Press.

247. Roche, A. F. & Himes, J. H. (1980). Incremental growth charts. *American Journal of Clinical Nutrition*, **33**, 2041–52.

248. Roche, A. F., Himes, J. H., Siervogel, R. M. & Johnson, D. L. (1979b). *Longitudinal Study of Human Hearing: Its Relationship to Noise and Other Factors. II. Results from the first three years*, AMRL-TR-79-102, pp. 221. Wright-Patterson Air Force Base, Ohio: Aerospace Medical Research Laboratory.

249. Roche, A. F. & Lewis, A. B. (1974). Sex differences in the elongation of the cranial base during pubescence. *Angle Orthodontist*, **44**, 279–94.

250. Roche, A. F. & Lewis, A. B. (1976). Late growth changes in the cranial base. In *Symposium on Development of the Basicranium*, ed. J. Bosma, pp. 221–36. Department of Health, Education and Welfare, Publication No. NIH 76-989, Bethesda, MD: US Government Printing Office.

251. Roche, A. F., Lewis, A. B., Wainer, H. & McCartin, R. (1977a). Late elongation of the cranial base. *Journal of Dental Research*, **56**, 802–7.

252. Roche, A. F. & Malina, R. M. (1983). *Manual of Physical Status and Performance, in Childhood, Vol. I. Physical Status*. New York: Plenum Press.

253. Roche, A. F. & McKigney, J. (1975). Physical growth of ethnic groups compromising the United States population. *American Journal of Clinical Nutrition*, **28**, 1071–4.

254. Roche, A. F. & Mukherjee, D. (1982). Ridge regressions to estimate body density and their serial errors. *American Journal of Physical Anthropology*, **57**, 221.

255. Roche, A. F., Mukherjee, D., Chumlea, W. C. & Champney, T. F. (1983a). Examination effects in audiometric testing of children. *Scandinavian Audiology*, **12**, 251–6.

256. Roche, A. F., Mukherjee, D., Chumlea, W. C. & Siervogel, R. M. (1983b). Iris pigmentation and AC thresholds. *Journal of Speech and Hearing Research*, **26**, 151–4.

257. Roche, A. F., Mukherjee, D. & Guo, S. (1986b). Head circumference growth patterns: birth to 18 years. *Human Biology*, **58**, 893–906.

258. Roche, A. F., Mukherjee, D., Guo, S. & Moore, W. M. (1987b). Head circumference reference data: birth to 18 years. *Pediatrics*, **79**, 706–12.

259. Roche, A. F., Mukherjee, D., Siervogel, R. M. & Chumlea, W. C. (1983e). Serial changes in auditory thresholds from 8 to 18 years in relation to environmental noise exposure. In *Noise as a Public Health Problem*, ed. G. Rossi, pp. 285–96. Milano, Italy: Centro Richerche e Studi Amplifon.

260. *Roche, A. F., Roberts, J. & Hamill, P. V. V. (1974c). *Skeletal Maturity of Children 6–11 Years: United States*. Vital and Health Statistics, Series 11, No. 140, Washington, DC: National Center for Health Statistics, Government Printing Office.
261. *Roche, A. F., Roberts, J. & Hamill, P. V. V. (1976a). *Skeletal Maturity of Youths 12–17 Years: United States*. Vital and Health Statistics, Series 11, No. 160, Washington, DC: National Center for Health Statistics, Government Printing Office.
262. Roche, A. F., Rogers, E. & Cronk, C. E. (1984). Serial analyses of fat-related variables. In *Human Growth and Development*, ed. J. Borms, R. Hauspie, A. Sand, C. Susanne & M. Hebbelinck, pp. 597–601. New York: Plenum Press.
263. Roche, A. F., Rohmann, C. G., French, N. Y. & Davila, G. H. (1970b). Effect of training on replicability of assessments of skeletal maturity (Greulich–Pyle). *American Journal of Roentgenology, Radium Therapy and Nuclear Medicine*, **108**, 511–15.
264. Roche, A. F., Siervogel, R. M., Chumlea, W. C., Reed, R. B., Valadian, I., Eichorn, D. & McCammon, R. W. (1982a). Serial changes in subcutaneous fat thicknesses of children and adults. *Monographs in Paediatrics*, **17**. Karger, Basel.
265. Roche, A. F., Siervogel, R. M., Chumlea, W. C. & Webb, P. (1981b). Grading body fatness from limited anthropometric data. *American Journal of Clinical Nutrition*, **34**, 2831–8.
266. Roche, A. F., Siervogel, R. M., Himes, J. H. & Johnson, D. L. (1977b). *Longitudinal Study of Human Hearing: Its Relationship to Noise and Other Factors. I. Design of Five Year Study; Data from First Year*. AMRL-TR-76-110, pp. 158. Wright-Patterson Air Force Base, Ohio: Aerospace Medical Research Laboratory.
267. Roche, A. F., Siervogel, R. M., Himes, J. H. & Johnson, D. L. (1978). Longitudinal study of hearing in children: Baseline data concerning auditory thresholds, noise exposure and biological factors. *Journal of the Acoustical Society of America*, **64**, 1593–601.
268. Roche, A. F., Siervogel, R. M. & Roche, E. M. (1979a). Digital dermatoglyphics in a white population from Southwestern Ohio. *Birth Defects: Original Article Series*, **15**, 389–409.
269. Roche, A. F., Tyleshevski, F. & Rogers, E. (1983c). Non-invasive measurement of physical maturity in children. *Research Quarterly for Sport and Exercise*, **54**, 364–71.
270. Roche, A. F., Wainer, H. & Thissen, D. (1974a). The RWT method for predicting adult stature. *Compte Rendu de La XIIᵉ Reunion des Equipes Chargees des Etudes sur la Croissance et le Developpement de l'enfant Normal*, pp. 50–62. Paris, France: Centre Internationale D'Enfance.
271. Roche, A. F., Wainer, H. & Thissen, D. (1974b). The prediction of growth in individual children. *Proceedings XIV International Congress in Pediatrics*, Buenos Aires, Argentina. No. 5005, pp. 1–11.
272. Roche, A. F., Wainer, H. & Thissen, D. (1975a). Predicting adult stature for individuals. *Monographs in Pediatrics*, **3**, 1–115.
273. Roche, A. F., Wainer, H. & Thissen, D. (1975b). The RWT method for the prediction of adult stature. *Pediatrics*, **56**, 1026–33.

## 270 References

274. Roche, A. F., Wainer, H. & Thissen, D. (1975d). *Skeletal Maturity: The Knee Joint as a Biological Indicator.* New York: Plenum Press.
275. Roche, E. M., Roche, A. F. & Siervogel, R. M. (1979c). Absence of triradius d in a three-generation pedigree and other variations of main-line d. *American Journal of Physical Anthropology*, **51**, 389–92.

Rohmann, C. G., Garn, S. M., Israel, H. & Ascoli, W. (1967). Continuing bone 'expansion' as a general phenomenon. *American Journal of Physical Anthropology*, **27**, 247.

Ross Laboratories. (1981). *Incremental Growth Charts.* Columbus, Ohio: Ross Laboratories.

Ross Laboratories. (1983). *Parent-specific Adjustments for Evaluation of Length and Stature.* Columbus, Ohio: Ross Laboratories.

*Scammon, R. E. (1927). The first seriatim study of human growth. *American Journal of Physical Anthropology*, **10**, 329–36.

Seegers, W. H. (1937a). The nitrogen balance of a young primipara. *American Journal of Obstetrics and Gynecology*, **34**, 1019–22.

Seegers, W. H. (1937b). The effect of protein deficiency on the course of pregnancy. *American Journal of Physiology*, **119**, 474–9.

Seegers, W. H. & Potgieter, M. (1937). The quantity of creatine and creatinine excreted in normal human pregnancy. *Human Biology*, **9**, 404–9.

Selby, S. (1961). Metaphyseal cortical defects in the tubular bones of growing children. *Journal of Bone and Joint Surgery*, **43-A**, 395–400.

Selby, S., Garn, S. M. & Kanareff, V. (1955). The incidence and familial nature of a bony bridge on the first cervical vertebra. *American Journal of Physical Anthropology*, **13**, 129–41.

Siervogel, R. M. (1983). Genetic and familial factors in essential hypertension and related traits. *Yearbook of Physical Anthropology*, **26**, 37–64.

Siervogel, R. M. (1984). Heredity of hypertension. In *National Heart, Lung and Blood Institute Workshop on Juvenile Hypertension*, ed. J. M. H. Loggie, M. J. Horan, A. B. Gruskin, A. R. Horn, J. B. Dunbar & R. F. Havlik, pp. 111–24. New York: Biomedical Information Corporation.

Siervogel, R. M. & Baumgartner, R. N. (1988). Fat distribution and blood pressures. In *Fat Distribution During Growth and Later Health Outcomes*, ed. C. Bouchard & F. E. Johnston, pp. 243–61. New York: Alan R. Liss.

Siervogel, R. M., Baumgartner, R. N., Roche, A. F., Chumlea, W. C. & Glueck, C. J. (1989b). Maturity and its relationship to plasma lipid and lipoprotein levels in adolescents. The Fels Longitudinal Study. *American Journal of Human Biology*, **1**, 217–26.

Siervogel, R. M., Guo, S. & Roche, A. F. (in press). Risk of adult overweight predicted from childhood values of body mass index. *Proceedings of the Fifth Conference for Federally Supported Human Nutrition Research Units and Centers*, Washington, DC.

Siervogel, R. M., Mukherjee, D. & Roche, A. F. (1984). Familial resemblance for patterns of change in weight/stature$^2$ (W/S$^2$) from 2 to 18 years. *American Journal of Human Genetics*, **36**, 180S.

Siervogel, R. M., Roche, A. F. & Chumlea, W. C. (1982a). Environmental sound exposure in children: its major sources and its effects on hearing and growth.

*Proceedings of the III International Congress of Auxology*, Brussels, Belgium, p. 129.

Siervogel, R. M., Roche, A. F., Chumlea, W. C., Morris, J. G., Webb, P. & Knittle, J. L. (1982c). Blood pressure, body composition, and fat tissue cellularity in adults. *Hypertension*, **4**, 382–6.

Siervogel, R. M., Roche, A. F., Guo, S., Mukherjee, D. & Chumlea, W. C. (1989a). Patterns of change in adiposity during childhood and their relation to adiposity at 18 years. *American Journal of Human Biology*, **1**, 136.

Siervogel, R. M., Roche, A. F., Guo, S., Mukherjee, D. & Chumlea, W. C. (in press). Patterns of change in weight/stature² from 2 to 18 years: Findings from long-term serial data for children in the Fels Longitudinal Growth Study. *International Journal of Obesity*.

Siervogel, R. M., Roche, A. F., Himes, J. H., Chumlea, W. C. & McCammon, R. (1982b). Subcutaneous fat distribution in males and females from 1 to 39 years of age. *American Journal of Clinical Nutrition*, **36**, 162–71.

Siervogel, R. M., Roche, A. F., Johnson, D. L. & Fairman, T. (1982d). Longitudinal study of hearing in children. II. Cross-sectional studies of noise exposure as measured by dosimetry. *Journal of the Acoustical Society of America*, **71**, 372–7.

Siervogel, R. M., Roche, A. F., Morris, J. G. & Glueck, C. J. (1981). Blood pressure and its relationship to plasma lipids and lipoproteins in children: Cross-sectional data from the Fels Longitudinal Study. *Preventive Medicine*, **10**, 555–63.

Siervogel, R. M., Roche, A. F. & Roche, E. M. (1978). Development fields for digital dermatoglyphic traits as revealed by multivariate analysis. *Human Biology*, **50**, 541–56.

Siervogel, R. M., Roche, A. F. & Roche, E. M. (1979). The identification of developmental fields using digital distributions of fingerprint patterns and ridge counts. *Birth Defects: Original Article Series*, **15**, 135–47.

Silverman, F. N. (1957). Roentgen standards for size of the pituitary fossa from infancy through adolescence. *American Journal of Roentgenology, Radium Therapy and Nuclear Medicine*, **78**, 451–60.

*Simmons, K. (1944). *The Brush Foundation Study of Child Growth and Development. II. Physical growth and development. Monographs of the Society for Research in Child Development*, **9**.

Smith, L., Rosner, B., Roche, A. & Guo, S. (1990). Modelling the correlation structure of blood pressure in the Fels data. *Annual Meeting of the American Statistical Association*, p. 137.

Smith, L. A., Rosner, B., Roche, A. F. & Guo, S. (in press). Serial changes in blood pressures from adolescence into adulthood. *American Journal of Epidemiology*.

Sobel, E. & Falkner, F. (1974). Normal and abnormal growth patterns of the newlyborn and the preadolescent. In *Endocrine and Genetic Disorders of Childhood and Adolescence*, 2nd edn, ed. L. I. Gardner, pp. 6–18. Philadelphia: W. B. Saunders.

276. Sontag, L. W. (1938). Evidences of disturbed prenatal and neonatal growth in bones of infants aged one month. *American Journal of Diseases of Children*, **55**, 1248–56.

277. Sontag, L. W. (1940). Effect of fetal activity on the nutritional state of the infant at birth. *American Journal of Diseases of Children*, **60**, 621–30.
278. Sontag, L. W. (1941). The significance of fetal environmental differences. *American Journal of Obstetrics and Gynecology*, **42**, 996–1003.
279. Sontag, L. W. (1944a). War and the fetal maternal relationship. *Marriage and Family Living*, **6**, 3–5.
280. Sontag, L. W. (1944b). Differences in modifiability of fetal behavior and physiology. *Psychosomatic Medicine*, **6**, 151–4.
281. Sontag, L. W. (1971). The history of longitudinal research: Implications for the future. *Child Development*, **42**, 987–1004.
282. Sontag, L. W. & Allen, J. E. (1947). Lung calcifications and histoplasmin-tuberculin skin sensitivity. *Journal of Pediatrics*, **30**, 657–67.
283. Sontag, L. W., Baker, C. T. & Nelson, V. L. (1958). *Mental growth and personality development: A longitudinal study. Monographs for the Society of Research in Child Development*, **23**.
284. Sontag, L. W. & Comstock, G. (1938). Striae in the bones of a set of monozygotic triplets. *American Journal of Diseases of Children*, **56**, 301–8.
285. Sontag, L. W. & Garn, S. M. (1954). Growth. *Annual Review of Physiology*, **16**, 37–50.
286. Sontag, L. W. & Garn, S. M. (1957). Human heredity studies of the Fels Research Institute. *Acta Genetica et Statistica Medica*, **61**, 494–502.
287. Sontag, L. W. & Harris, L. M. (1938). Evidences of disturbed prenatal and neonatal growth in bones of infants aged one month: II. Contributing factors. *American Journal of Diseases of Children*, **56**, 1248–55.
288. Sontag, L. W. & Lipford, J. (1943). The effects of illness and other factors on the appearance pattern of skeletal epiphyses. *Journal of Pediatrics*, **23**, 391–409.
289. Sontag, L. W., Munson, P. & Huff, E. (1936). Effects on the fetus of hypervitaminosis D and calcium and phosphorus deficiency during pregnancy. *American Journal of Diseases of Children*, **51**, 302–10.
290. Sontag, L. W. & Nelson, V. L. (1933). A study of identical triplets. Part I. Comparison of the physical and mental traits of a set of monozygotic dichorionic triplets. *Journal of Heredity*, **24**, 473–80.
291. Sontag, L. W. & Newbery, H. (1940). Normal variations of fetal heart rate during pregnancy. *American Journal of Obstetrics and Gynecology*, **40**, 449–52.
292. Sontag, L. W. & Newbery, H. (1941). Incidence and nature of fetal arrhythmias. *American Journal of Diseases of Children*, **62**, 991–9.
293. Sontag, L. W. & Potgieter, M. (1938). The variability of nitrogen excretion in twenty-four hour periods as compared with that in longer periods. *Human Biology*, **10**, 400–8.
294. Sontag, L. W. & Pyle, S. I. (1941a). Variations in the calcification pattern in epiphyses. *American Journal of Roentgenology, Radium Therapy, and Nuclear Medicine*, **45**, 50–4.
295. Sontag, L. W. & Pyle, S. I. (1941b). The appearance and nature of cyst-like areas in the distal femoral metaphyses of children. *American Journal of Roentgenology, Radium Therapy, and Nuclear Medicine*, **46**, 185–8.
296. Sontag, L. W., Pyle, S. I. & Cape, J. (1935). Prenatal conditions and the status of infants at birth. *American Journal of Diseases of Children*, **50**, 337–42.

297. Sontag, L. W. & Reynolds, E. L. (1944). Ossification sequences in identical triplets. A longitudinal study of resemblances and differences in the ossification patterns of a set of monozygotic triplets. *Journal of Heredity*, **35**, 57–64.
298. Sontag, L. W. & Reynolds, E. L. (1945). The Fels Composite Sheet. I: A practical method for analyzing growth progress. *Journal of Pediatrics*, **26**, 327–35.
299. Sontag, L. W., Reynolds, E. L. & Torbet, V. (1944). The relations of basal metabolic gain during pregnancy to non-pregnant basal metabolism. *American Journal of Obstetrics and Gynecology*, **48**, 315–20.
300. Sontag, L. W. & Richards, T. W. (1938). *Studies in fetal behavior: I. Fetal heart rate as a behavioral indicator. Monograph of Society for Research in Child Development* **3**.
301. Sontag, L. W., Seegers, W. H. & Hulstone, L. (1938). Dietary habits during pregnancy (with special reference to the value of qualitative food records). *American Journal of Obstetrics and Gynecology*, **35**, 614–21.
302. Sontag, L. W., Snell, D. & Anderson, M. (1939). Rate of appearance of ossification centers from birth to the age of five years. *American Journal of Diseases of Children*, **58**, 949–56.
303. Sontag, L. W. & Wallace, R. F. (1933). An apparatus for recording fetal movement. *American Journal of Psychology*, **55**, 517–19.
304. Sontag, L. W. & Wallace, R. F. (1934). Preliminary report of the Fels Fund. Study of fetal activity. *American Journal of Diseases of Children*, **48**, 1050–7.
305. Sontag, L. W. & Wallace, R. F. (1935a). The effect of cigarette smoking during pregnancy upon the fetal heart rate. *American Journal of Obstetrics and Gynecology*, **29**, 77–82.
306. Sontag, L. W. & Wallace, R. F. (1935b). The movement response of the human fetus to sound stimuli. *Child Development*, **6**, 253–8.
307. Sontag, L. W. & Wallace, R. F. (1936). Changes in the rate of the human fetal heart in response to vibratory stimuli. *American Journal of Diseases of Children*, **51**, 583–9.
308. Sontag, L. W. & Wines, J. (1947). Relation of mothers' diets to status of their infants at birth and in infancy. *American Journal of Obstetrics and Gynecology*, **54**, 994–1003.
Spence, M. A., Falk, C. T., Neiswanger, K., Field, L. L., Marazita, M. L., Allen, F. H., Jr., Siervogel, R. M., Roche, A. F., Crandal, B. F. & Sparks, R. S. (1984). Estimating the recombination frequency for the PTC–Kell linkage. *Human Genetics*, **67**, 183–6.
Spencer, R. P. & Coulombe, M. J. (1966). Quantification of the radiographically determined age dependence of bone thickness. *Investigative Radiology*, **1**, 144–7.
Spencer, R. P., Garn, S. M. & Coulombe, M. J. (1966). Age-dependent changes in metacarpal cortical thickness in two populations. *Investigative Radiology*, **1**, 394–7.
Spencer, R. P., Sagel, S. S. & Garn, S. M. (1968). Age changes in five parameters of metacarpal growth. *Investigative Radiology*, **3**, 27–34.
Sykes, R. C. (1985). Secular Change in Highly Heritable Traits: IQ and Stature. Doctoral Dissertation, Department of Behavioral Science, Chicago, Illinois: The University of Chicago.

Thissen, D. (1989). Statistical estimation of skeletal maturity. *American Journal of Human Biology*, **1**, 185–92.

Thissen, D., Bock, R. D., Wainer, H. & Roche, A. F. (1976). Individual growth in stature: Comparison of four U.S. growth studies. *Annals of Human Biology*, **3**, 527–42.

Thomas, C. B. & Garn, S. M. (1960). Degree of obesity and serum cholesterol level. *Science*, **131**, 42.

Townsend, E. (1978). Comparison of a new skinfold caliper to Lange and Harpenden calipers. *American Journal of Physical Anthropology*, **48**, 443.

*Tukey, J. W. (1972). *Exploratory Data Analysis*, 2nd preliminary edn. Reading: Addison-Wesley.

Wainer, H., Roche, A. F. & Bell, S. (1978). Predicting adult stature without skeletal age and without paternal data. *Pediatrics*, **61**, 569–72.

Wainer, H. & Thissen, D. (1975). Multivariate semi-metric smoothing in multiple prediction. *Journal of the American Statistical Association*, **70**, 568–73.

Welford, N., Sontag, L. W., Phillips, W. & Phillips, D. (1967). Individual differences in heart rate variability in the human fetus. *American Journal of Obstetrics and Gynecology*, **98**, 56–61.

*White House Conference on Child Health and Protection, Section I. (1933). *Growth and Development of the Child*. Part II. *Anatomy and Physiology*, pp. 1–629. New York: Century.

Wolański, N. (1966a). A new method for the evaluation of tooth formation. *Acta Genetica*, **16**, 186–97.

Wolański, N. (1966b). The interrelationship between bone density and cortical thickness in the second metacarpal as a function of age. In *Progress in Development of Methods in Bone Densitometry*, pp. 65–77. NASA SP-64. Washington, DC: National Aeronautics and Space Administration.

Wolański, N. (1967). New method for the evaluation of tooth formation. *Journal of Dental Research*, **46**, 875.

Woynarowska, B., Mukherjee, D., Roche, A. F. & Siervogel, R. M. (1985). Blood pressure changes during adolescence and subsequent adult blood pressure level. *Hypertension*, **7**, 695–701.

Xi, H. & Roche, A. F. (1990). Differences between the hand–wrist and the knee in assigned skeletal ages. *American Journal of Physical Anthropology*, **83**, 95–102.

Xi, H., Roche, A. F. & Baumgartner, R. N. (1989a). Association of adipose tissue distribution with relative skeletal age in boys: The Fels Longitudinal Study. *American Journal of Human Biology*, **1**, 589–96.

Xi, H., Roche, A. F. & Guo, S. (1989b). Sibling correlations for skeletal age assessments by the FELS method. *American Journal of Human Biology*, **1**, 613–19.

Yarbrough, C., Habicht, J. P., Klein, R. E. & Roche, A. F. (1973). Determining the biological age of the preschool child from a hand–wrist radiograph. *Investigative Radiology*, **8**, 233–43.

Young, R. W. (1956). The measurement of cranial shape. *American Journal of Physical Anthropology*, **14**, 59–71.

Young, R. W. (1957). Postnatal growth of the frontal and parietal bones in white males. *American Journal of Physical Anthropology*, **15**, 367–86.

Young, R. W. (1959). Age changes in the thickness of the scalp in white males. *Human Biology*, **31**, 74–9.

# Index

tracking (*cont.*)
  adipose tissue thickness, 226–8
  definitions, 226
  plasma lipids, 240
transformation of variables, 40–1
triceps adipose tissue thickness
  and peak height velocity, 218
  tracking, 227
trunk circumference and pelvic growth,
    180

ulnar sesamoid ossification
  and cranial base length, 167–8
  and mandibular growth, 176–7
urinary creatinine and creatine, 55, 209–10

vertebrae
  enlargement of third cervical in adults,
    177
  first cervical bridge, 184
    familial effects, 184
  lumbo–sacral fusion, 184–5
    familial effects, 185–7

weight
  and adipose tissue distribution, 223
  and blood pressure, 237
  and peak height velocity, 106–7
  and pelvic growth, 179
  at menarche, 147

growth after menarche, 106–7
growth patterns, 72–4
relative, 205
risk analysis, 208
weight-for-recumbent length, 85
weight-for-stature, 40, 85–6
weight/recumbent length$^3$ and pelvic
    growth, 179
weight/stature$^2$, 39–40, 85–6
  age changes, 204–8
  and nutritional status, 231–2
  changes and adult status, 207
  changes in adults, 204
  changes in pubescence, 208
  curve fitting, 207
weight/stature$^3$ and pelvic growth, 179
weight–stature index
  development of, 206
  patterns of change, 207–8
  risk analysis, 208
  tracking, 205–7
Wright State University, 4–5

X-chromosome effects
  cortical thickness, 189
  dental development, 62–3
  missing ossification centers, 183
  onset of ossification, 126

zygosity diagnosis, 65–6